MICROBIOLOGICAL RISK ASSESSMENT SERIES

16

Risk assessment of *Vibrio parahaemolyticus* in seafood

INTERPRETATIVE SUMMARY AND TECHNICAL REPORT

WORLD HEALTH ORGANIZATION

FOOD AND AGRICULTURE ORGANIZATION OF THE UNITED NATIONS

2011

WHO Library Cataloguing-in-Publication Data
Risk assessment of *Vibrio parahaemolyticus* in seafood: interpretative summary and technical report.
(Microbiological risk assessment series no. 16)
1. *Vibrio parahaemolyticus.* 2.Seafood. 3.Food microbiology. 4.Food contamination.
5.Risk assessment. I.World Health Organization. II.Food and Agriculture Organization of the United Nations. III.Series.

ISBN 978-92-4-154817-5 (WHO) (NLM classification: QW 85)
ISBN 978-92-5-106874-8 (FAO)
ISSN 1726-5274

Recommended citation
FAO/WHO [Food and Agriculture Organization of the United Nations/World Health Organization]. 2011. Risk assessment of *Vibrio parahaemolyticus* in seafood: Interpretative summary and Technical report. Microbiological Risk Assessment Series No. 16. Rome. 193pp

Contents

INTERPRETATIVE SUMMARY

TECHNICAL REPORT

Part I – Microbiological risk assessment of *Vibrio parahaemolyticus* in raw oysters

Part II – Microbiological risk assessment of *Vibrio parahaemolyticus* in *Anadara granosa* (Bloody clam)

Part III – Microbiological risk assessment of *Vibrio parahaemolyticus* in finfish

FOREWORD

The Members of the Food and Agriculture Organization of the United Nations (FAO) and of the World Health Organization (WHO) have expressed concern regarding the level of safety of food at both national and international levels. Increasing foodborne disease incidence over the last decades seems, in many countries, to be related to an increase in disease caused by micro-organisms in food. This concern has been voiced in meetings of the Governing Bodies of both Organizations, and in the Codex Alimentarius Commission. It is not easy to decide whether the suggested increase is real or an artefact of changes in other areas, such as improved disease surveillance or better detection methods for micro-organisms in foods. However, the important issue is whether new tools or revised and improved actions can contribute to our ability to lower the disease burden and provide safer food. Fortunately new tools, which can facilitate actions, seem to be on their way.

Over the past decade, Risk Analysis—a process consisting of risk assessment, risk management and risk communication—has emerged as a structured model for improving our food control systems, with the objectives of producing safer food, reducing the numbers of foodborne illnesses and facilitating domestic and international trade in food. Furthermore, we are moving towards a more holistic approach to food safety, where the entire food chain needs to be considered in efforts to produce safer food. As with any model, tools are needed for the implementation of the risk analysis paradigm.

Risk assessment is the science-based component of risk analysis. Science today provides us with in-depth information on life in the world we live in. It has allowed us to accumulate a wealth of knowledge on microscopic organisms, their growth, survival and death, even their genetic make-up. It has given us an understanding of food production, processing and preservation, and of the link between the microscopic and the macroscopic world and how we can benefit from as well as suffer from these micro-organisms. Risk assessment provides us with a framework for organizing all this data and information and to better understand the interaction between micro-organisms, foods and human illness. It provides us with the ability to estimate the risk to human health from specific micro-organisms in foods and gives us a tool with which we can compare and evaluate different scenarios, as well as to identify the types of data necessary for estimating and optimizing mitigating interventions.

Microbiological risk assessment can be considered as a tool that can be used in the management of the risks posed by foodborne pathogens and in the elaboration of standards for food in international trade. However, undertaking a microbiological risk assessment (MRA), particularly quantitative MRA, is recognized as a resource-intensive task requiring a multidisciplinary approach. Yet foodborne illness is among the most widespread public health problems, creating social and economic burdens as well as human suffering, making it a concern that all countries need to address. As risk assessment can also be used to justify the introduction of more stringent standards for imported foods, a knowledge of MRA is important for trade purposes, and there is a need to provide countries with the tools for understanding and, if possible, undertaking MRA. This need, combined with that of the Codex Alimentarius for risk-based scientific advice, led FAO and WHO to undertake a programme of activities on MRA at the international level. The Nutrition and Consumer Protection Division, FAO, and the Department of Food Safety and Zoonoses, WHO, are the lead units responsible for this initiative. The two groups have worked together to develop the area of MRA at the international level for application at both the national and international levels.

This work has been greatly facilitated by the contribution of people from around the world with expertise in microbiology, mathematical modelling, epidemiology and food technology, to name but a few. This Microbiological Risk Assessment series provides a range of data and information to those who need to understand or undertake MRA. It comprises risk assessments of particular pathogen-commodity combinations, interpretative summaries of the risk assessments, guidelines for undertaking and using risk assessment, and reports addressing other pertinent aspects of MRA.

We hope that this series will provide a greater insight into MRA, how it is undertaken and how it can be used. We strongly believe that this is an area that should be developed in the international sphere, and have already from the present work clear indications that an international approach and early agreement in this area will strengthen the future potential for use of this tool in all parts of the world, as well as in international standard setting. We would welcome comments and feedback on any of the documents within this series so that we can endeavour to provide Member countries, Codex Alimentarius and other users of this material with the information they need to use risk-based tools, with the ultimate objective of ensuring that safe food is available for all consumers.

Mr Samuel C. Jutzi

Officer-in-Charge
Nutrition and Consumer Protection Division
Food and Agriculture Organization of the United Nations

Dr M. Maged Younes

Director
Department of Food Safety and Zoonoses
World Health Organization

ACKNOWLEDGEMENTS

The Food and Agriculture Organization of the United Nations and the World Health Organization would like to express their appreciation to all those who contributed to the preparation of this report. Special appreciation is extended to the risk assessment drafting group on *Vibrio* spp. in seafood. We also wish to express deepest gratitude to those additional contributors to Part II – Bloody clams, in particular, Dr Varaporn Vuddhakul's group in the Faculty of Science and Dr Sineenart Kalnauwakul's group in the Faculty of Medicine at Prince of Songkla University, Thailand. Many people provided their time and expertise by reviewing the report and providing their comments, additional data and other relevant information. Both the risk assessment drafting group and the reviewers are listed in the following pages.

Appreciation is also extended to all those who responded to the calls for data that were issued by FAO and WHO and brought to our attention data not readily available in the mainstream literature and official documentation.

The risk assessment work was coordinated by the Secretariat of the Joint FAO/WHO Expert Meetings on Microbiological Risk Assessment (JEMRA). This included Sarah Cahill, Maria de Lourdes Costarrica and Jean-Louis Jouve (2001–2004) in FAO, and Peter Karim Ben Embarek, Jocelyne Rocourt (2001–2004), Hajime Toyofuku (1999–2004), Jørgen Schlundt (2000–2010) and Kazuko Fukushima in WHO. During the development of the risk assessment, additional support and feedback were provided by Lahsen Ababouch, Henri Loreal, Hector Lupin and Iddya Karunasagar, Fishery Industries Division in FAO, and Jeronimas Maskeliunas, Codex Secretariat. Sarah Cahill coordinated the finalization and publication of the report. Thorgeir Lawrence was responsible for editing the report for language and preparation for printing.

The preparatory work and the publication of this report was supported and funded by the FAO Nutrition and Consumer Protection Division, the FAO Fisheries and Aquaculture Department and the WHO Department of Food Safety and Zoonoses.

Risk Assessment Drafting Group

John BOWERS	Department of Health and Human Services, United States Food and Drug Administration, United States of America
Anders DALSGAARD	Department of Veterinary Microbiology, The Royal Veterinary and Agricultural University, Denmark
Angelo DEPAOLA	Office of Seafood, United States Food and Drug Administration, United States of America
Indriani KARUNASAGAR	Department of Fishery Microbiology, University of Agricultural Sciences, India
Thomas MCMEEKIN	Australian Food Safety Centre of Excellence, University of Tasmania, Australia
Mitsuaki NISHIBUCHI	Center for Southeast Asian Studies, Kyoto University, Japan
Ken OSAKA	National Institute of Infectious Diseases, Ministry of Health, Labour and Welfare, Japan
John SUMNER	M&S Food Consultants Pty. Ltd., Australia
Mark WALDERHAUG	Center for Biologics Evaluation and Research, United States Food and Drug Administration, United States of America

Reviewers (Participants in expert consultations)

Awa Kane AÏDARA	Institut Pasteur de Dakar, Laboratoire de Bactériologie Expérimentale, Senegal
Nourredine BOUCHRITI	Department d'Hygiéne et d'Industrie des Denrées Alimentaires d'Origine Animale, Institut Agronomique et Vétérinaire Hassan II, Morocco
Bjarke Bak CHRISTENSEN	Danish Veterinary and Food Administration, Denmark
John COWDEN	Scottish Centre for Infection and Environmental Health, United Kingdom
Louis Anthony COX	Cox Associates, United States of America
Aamir FAZIL	Population and Public Health Branch, Health Canada, Canada
Heriberto FERNÁNDEZ	Instituto de Microbiología Clínica, Universidad Austral de Chile, Chile
Jean-Michel FOURNIER	Centre National de Référence des Vibrions et du Choléra, Institut Pasteur, France
Marja-Liisa HÄNNINEN	Department of Food and Environmental Hygiene, Faculty of Veterinary Medicine, Helsinki University, Finland
Emma HARTNETT	Department of Risk Research, Veterinary Laboratories Agency (Weybridge), United Kingdom
Tom HUMPHREY	Department of Clinical Veterinary Science, University of Bristol, United Kingdom
Susana María DE LOS MILAGROS JIMÉNEZ	Departamento de Microbiología, Universidad Nacional del Litoral, Argentina
Anna LAMMERDING	Population and Public Health Branch, Health Canada, Canada
Ron LEE	Centre for Environment, Fisheries & Aquaculture Science (Cefas), Weymouth Laboratory, United Kingdom
Carlos LIMA DOS SANTOS	Private Consultant, Brazil
Dorothy-Jean MCCOUBREY	Ministry of Agriculture and Forestry, New Zealand

Geoffrey MEAD	Private Consultant, United Kingdom
Paul MEAD	Foodborne and Diarrheal Diseases Branch, Centers for Disease Control and Prevention, United States of America
Marianne MILIOTIS	Food and Drug Administration, Center for Food Safety and Applied Nutrition (CFSAN), United States of America
Noel MURRAY	Biosecurity Authority, Ministry of Agriculture and Forestry, New Zealand
George NASINYAMA	Department of Epidemiology and Food Safety, Faculty of Veterinary Medicine, Makerere University, Uganda
Maarten NAUTA	National Institute for Public Health and the Environment (RIVM), The Netherlands
Diane G. NEWELL	Veterinary Laboratories Agency (Weybridge), United Kingdom
Mitsuaki NISHIBUCHI	Center for Southeast Asian Studies, Kyoto University, Japan
Servé NOTERMANS	TNO Nutrition and Food Research Institute, The Netherlands
Greg PAOLI	Decisionalysis Risk Consultants, Inc., Canada
Pensri RODMA	Department of Medical Science, Ministry of Public Health, Thailand
Hanne ROSENQUIST	Danish Veterinary and Food Administration, Denmark
Sasitorn KANARAT	Veterinary Public Health Laboratory, Department of Livestock Development, Thailand
Mark TAMPLIN	Australian Food Safety Centre of Excellence, University of Tasmania, Australia
Paul VANDERLINDE	Food Science Australia, Australia
Henrik WEGENER	Danish Zoonosis Centre, Danish Veterinary Laboratory, Denmark
Shigeki YAMAMOTO	National Institute of Infectious Diseases, Ministry of Health Labour and Welfare, Japan

Peer reviewers

Swapan K. BANERJEE	Food Directorate, Health Canada, Ottawa, Canada
Enrico BUENAVENTURA	Canadian Food Inspection Agency, Food Safety Directorate, Ottawa, Canada
Phil BUSBY	New Zealand Food Safety Authority, Wellington, New Zealand
Ben DAUGHTRY	South Australian Research and Development Institute, Adelaide, Australia
Dalia dos PRAZERES RODRIGUES	Instituto Oswaldo Criz, Brazil
Romilio T. ESPEJO	Instituto de Nutrición y Tecnología de los Alimentos. Universidad de Chile, Chile
William R. JONES	Division of Seafood Safety, Office of Food Safety, CFSAN/USFDA, Washington DC, United States of America
Ron LEE	Centre for Environment Fisheries and Aquaculture Science, Weymouth, United Kingdom
Thomas MADIGAN	South Australian Research Development Institute, Adelaide, Australia
Dorothy-Jean MCCOUBREY	New Zealand Food Safety Authority, Auckland, New Zealand
Mitsuaki NISHIBUCHI	Center for Southeast Asian Studies, Kyoto University, Japan
Andrew POINTON	South Australian Research Development Institute, Adelaide, Australia
Marie-Laure QUILICI	Centre National de Référence des vibrions et du choléra, Institut Pasteur, France
T. RAMAMURTHY	National Institute of Cholera and Enteric Diseases, Kolkata, India
Rachel RANGDALE	Centre for Environment Fisheries and Aquaculture Science, Weymouth, United Kingdom
Alan REILLY	Food Safety Authority of Ireland, Dublin, Ireland

Son RADU	Department of Food Science, University Putra Malaysia, Serdang, Selangor, Malaysia
Irma Nelly GUTIERREZ RIVERA	Universidade de S. Paulo, S. Paulo, Brazil
Tom ROSS	Australian Food Safety Centre of Excellence, Hobart, Tasmania, Australia
Md. SIRAJUL ISLAM	International Centre for Diarrhoeal Disease Research, Bangladesh
Mark TAMPLIN	Australian Food Safety Centre of Excellence, Hobart, Tasmania, Australia
Donald SCHAFFNER	Rutgers University, United States of America
Connor THOMAS	School of Molecular and Biomedical Science, University of Adelaide, Adelaide, South Australia, Australia
Kaye WACHSMUTH	International public health consultant. 144 North Stone Street, DeLand, FL 32720, United States of America

Abbreviations used in the text

CCFH	Codex Committee on Food Hygiene
CDC	Centers for Disease Control and Prevention [United States of America]
Cefas	Centre for Environment, Fisheries & Aquaculture Science [United Kingdom]
cfu	colony forming unit
CI	Confidence Interval
FDA	Food and Drug Administration [United States of America]
FDA-VPRA	*Vibrio parahaemolyticus* Risk Assessment [of the FDA]
g	gram
GCSL	FDA Gulf Coast Seafood Laboratory, Dauphin Island
GS-PCR	group-specific polymerase chain reaction
h	hour(s)
HGMF	Hydrophobic Grid Membrane Filtration procedure
ISSC	Interstate Shellfish Sanitation Conference [United States of America]
KP+	Kanagawa-positive
min	minute(s)
mL	millilitre(s)
MLE	maximum likelihood estimates
MPN	most probable number
NBDC	National Buoy Data Center [United States of America]
NOAA	National Oceanic and Atmospheric Administration [United States of America]
NSSP	National Shellfish Sanitation Program [United States of America]
NSW	New South Wales [Australia]
PAC	polyaluminum chloride (as in disinfection by PAC coagulation)
PCR	polymerase chain reaction
PFGE	Pulsed Field Gel Electrophoresis
ppt	parts per thousand
TDH	Thermostable direct haemolysin
tdh	Thermostable direct haemolysin gene
TRH	TDH-related haemolysin
trh	TDH-related haemolysin gene
VBNC	viable but non-culturable

INTERPRETATIVE SUMMARY

Introduction [1]

Risk assessments are resource-intensive, data-driven activities designed to provide risk managers with the ability to consider a range of mitigations that are intended to improve public health. Because of the commitment of resources required to obtain useful models of a food production system, it is a benefit to risk managers when risk assessments are developed that have the potential to be generalized to other organisms, other commodities, and other national food systems. The present work uses an oyster harvest public health model developed in one country to assess risk in oysters from harvesting areas in other countries. The approach taken is then applied to other products, Bloody clam and finfish to determine to what extent such risk assessments can be adapted. This summary thus highlights the main findings of three pieces of work: risk assessment of *Vibrio parahaemolyticus* in raw oysters; risk assessment of *V. parahaemolyticus* in Bloody clam; and risk assessment of *V. parahaemolyticus* in finfish consumed raw.

Statement of purpose

The purpose of the risk assessment work on *Vibrio parahaemolyticus* in raw oysters was to use the model developed during the U.S. Quantitative Risk Assessment on the Public Health Impact of Pathogenic *Vibrio parahaemolyticus* in Raw Oysters (FDA-VPRA) (FDA, 2005) to estimate risk of illness from this pathogen due to consumption of oysters in Australia, Canada, Japan and New Zealand. The scientific underpinning for the present study has been published in FDA (2005) and can be accessed at http://www.fda.gov/Food/ScienceResearch/ResearchAreas/RiskAssessmentSafetyAssessment/ucm050421.htm.

The Bloody clam risk assessment was performed as a case study of performing a quantitative risk assessment in a developing country using local data. It focused on one city in Thailand, where consumption of Bloody clam is popular and where there is a public perception that this mollusc is a major source of diarrhoeal illness, including that caused by *V. parahaemolyticus*.

The purpose of the finfish risk assessment was to estimate the risk of *V. parahaemolyticus* infection from raw horse mackerel consumption in Japan, and to estimate the risk reduction from washing this fish after harvest or during preparation.

Constraints

When the study for raw oysters was initiated it was hoped that country-specific data would be supplied by a number of countries. In the event, only Australia, Canada, Japan and New Zealand were able to participate, and were able to supply only some of the data needed to run the model. Where country-specific data were lacking,

1. Note from editor: to maintain conformity between the Interpretative Summary and the three main Parts of the Technical Report, the table and figure numbers used here are those used in the respective Part.

United States of America surrogate data were used, placing a major constraint on this project (Table I–26).

Table I–26. Data sources for the *V. parahaemolyticus* (Vp) model.

	Australia	Canada	Japan	New Zealand
Water temperature	Local data	USA Pacific Northwest data	Local data	Local data
Relationship of water temperature and Vp levels	USA data	USA data	USA data	USA data
Time oysters are out of the water	Local harvest data	Local harvest data	Local harvest data	Local harvest data
Air temperature	Local data	Local data	Local data	Local data
Growth rate adjustment factor	USA data	USA data	USA data	USA data
Time oysters are refrigerated	Local data	Local data	Local data	Local data
Weight of oyster	USA data	USA data	USA data	USA data
Number of oysters consumed at one meal	USA data	USA data	USA data	USA data
Fraction of Vp that are *tdh*+ or *trh*+	USA data	USA data	USA data	USA data
Amount of oysters harvested from location	Local data	Local data	Local data	Local data
Multiplier for under-reporting	USA data	USA data	USA data	USA data
Fraction of oysters eaten raw	USA data	USA data	USA data	USA data

The Bloody clam risk assessment is based on limited data for prevalence of total and pathogenic (*tdh*+ and/or *trh*+) *V. parahaemolyticus* in thirty-two sample sets at harvest and retail stage. Seasonal variation, effect of environmental factors on prevalence and growth during harvest to consumption pathway could not be modelled due to lack of data. This lack of data is a challenge faced by many, particularly developing, countries in terms of trying to apply risk assessment. Risk estimates were based on consumption of cooked clams and the assessment used the FDA-VPRA dose response model, which is for *tdh*+ strains. The fraction of Bloody clam meals that were inadequately cooked was estimated based on a small survey. Other methods of preparation before consumption which are known to be practiced, such as grilling and frying, could not be modelled.

The finfish risk assessment is also based on limited data for the prevalence and concentration of *V. parahaemolyticus* in horse mackerel at different steps in the food chain. Differences in prevalence in various parts of the fish, such as gills, intestine or surface, could not be considered. Effect of environmental factors like temperature and salinity on prevalence, and time × temperature effects between harvest, landing and transport on *V. parahaemolyticus* prevalence were considered in the model. The estimation of the proportion of pathogenic strains (*tdh*+ and *trh*+) was based on published data for other seafoods, while estimation of growth during the post-harvest stage was based on the FDA-VPRA model for growth in oysters. Cross-contamination during preparation was not modelled and risk characterization was based on the FDA-VPRA dose response model for *tdh*+ strains.

Hazard identification

Vibrio parahaemolyticus is a marine micro-organism native in estuarine waters throughout the world. The organism was first identified as a foodborne pathogen in Japan in the 1950s (Fujino et al., 1953). By the late 1960s and early 1970s, *V. parahaemolyticus* was recognized as a cause of diarrhoeal disease worldwide, although most common in Asia and the United States of America. Vibrios concentrate in the gut of filter-feeding molluscan shellfish, such as oysters, clams and mussels, where they multiply and cohere. Although thorough cooking destroys these organisms, oysters are often eaten raw and, at least in the United States of America, are the most common food associated with *V. parahaemolyticus* infection (Hlady, 1997). Early studies in Japan showed that 96% of clinical strains produce a thermostable direct haemolysin (TDH), while only 1% of the environmental strains produce this haemolysin (Sakazaki, Iwanami and Tamura, 1968). Subsequently, TDH negative strains from clinical cases were found to produce a TDH-related haemolysin, TRH (Honda, Ni and Miwatani, 1988). At present, strains producing TDH and TRH are considered pathogenic to man. Diverse serotypes may be associated with human infections, but, recently, strains belonging to the O3:K6 serotype have been found to be the causative agent of several outbreaks in different countries (Nair et al., 2007).

In Asia, *V. parahaemolyticus* is a common cause of foodborne disease. In general, the outbreaks are small in scale, involving fewer than 10 cases, but occur frequently. Prior to 1994, the incidence of *V. parahaemolyticus* infections in Japan had been declining; however, there were 1280 reports of infection due to the organism in 1994–95 (Anon., 1999a) and during this period *V. parahaemolyticus* food poisonings outnumbered those of *Salmonella* food poisoning. From 1996 to 1998 there were 496 outbreaks and 24 373 cases of *V. parahaemolyticus* reported, while 25 211 cases were reported from 1999 to 2005 (Ministry of Health Labour and Welfare data). In general, outbreaks were more prevalent in the summer, with a peak in August.

In Taiwan, between 1986 and 1995, some 197 outbreaks of foodborne disease were caused by *V. parahaemolyticus* (Pan et al., 1997), while over 200 outbreaks were reported in 1997, including an outbreak of 146 cases acquired from boxed lunches (Anon., 1999b).

During 1997 and 1998 there were more than 700 cases of illness due to *V. parahaemolyticus* in the United States of America, the majority of which were associated with the consumption of raw oysters. In two of the 1998 outbreaks a serotype of *V. parahaemolyticus* (O3:K6), reported previously only in Asia, emerged for the first time as a principal cause of illness. Subsequent studies on these strains have revealed their pandemic spread. It was suggested that water temperatures warmer than usual were responsible for the outbreaks.

Though there is no clear epidemiological link between consumption of Bloody clam and illness due to *V. parahaemolyticus*, 38 of the 80 residents interviewed in Hat Yi city in Thailand reported experiencing diarrhoea after consumption of Bloody clam. During 1999, *V. parahaemolyticus* was isolated from 319 of 11 474 diarrhoeal stool samples from two hospitals in this city, with peak isolations during summer.

Annually, 500 to 800 *V. parahaemolyticus* outbreaks affecting about 10 000 people are reported in Japan, and *sashimi* (pieces of raw fish fillet) and *sushi* (vinegary rice ball with raw fish fillet) are responsible for 26% and 23% of outbreaks, respectively. Implicated seafood is rarely identified, but three outbreaks in 1999 were associated with horse mackerel. Mackerel was involved in 12 of 51 outbreaks investigated in Thailand.

In recent years there have also been significant outbreaks of *V. parahaemolyticus* infections in South America. Chile in particular has been affected in recent years, with more than 10 000 cases linked to mussel consumption during 2004 and 2005.

Few data exist in Europe on the incidence of *V. parahaemolyticus* infections, one reason being that such infections are not notifiable. However, the current knowledge of the incidence in Europe has been presented in 'Opinion of the Scientific Committee on Veterinary Measures relating to Public Health on *Vibrio vulnificus* and *Vibrio parahaemolyticus* in raw and undercooked seafood' (European Commission, 2001).

Exposure Assessments

Exposure assessment for raw oysters

The exposure assessment for raw oysters quantified exposure of consumers to pathogenic *V. parahaemolyticus* from the consumption of raw oysters in the United States of America, Canada, Japan, New Zealand and Australia, using the model developed for the FDA-VPRA (FDA, 2005). The model incorporates all phases in the harvest–post-harvest–consumption continuum, using a modular approach. The country-specific data that were available for the four selected areas are noted in Table I–26 above

Four factors were used to model exposure:

- level of pathogenic *V. parahaemolyticus* in seafood at harvest;
- effect of post-harvest handling and processing;
- ability of the organism to multiply to an infective dose; and
- number of pathogenic *V. parahaemolyticus* consumed.

The exposure assessment was divided into two separate modules: a Harvest Module; and a Post-harvest Module that incorporated retail and consumption sections. The elements of the Harvest Module are illustrated in Figure I–2. The main factors that potentially affect numbers of pathogenic *V. parahaemolyticus* in oysters at time of harvest are water temperature and salinity, though, since preliminary modelling demonstrated that salinity is not as strong as water temperature, it is represented as a dotted bubble in Figure I–2.

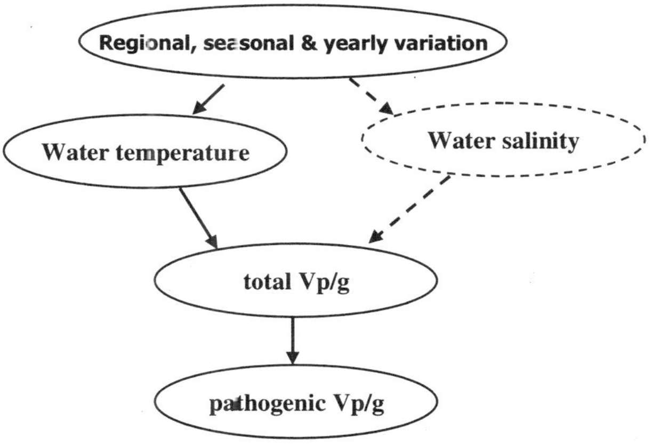

Figure I–2. Schematic depiction of the Harvest Module of the *V. parahaemolyticus* (Vp) risk assessment model.

The Post-harvest Module describes the effects of typical industry practices, including transportation, handling and processing, as well as distribution, storage and retail, on *V. parahaemolyticus* numbers in oysters harvested from various locations and in different seasons. Factors considered as influencing the numbers of pathogenic *V. parahaemolyticus* at consumption include:

- ambient air temperatures at time of harvest;
- time from harvest until oysters are placed under refrigeration;
- time it takes for oysters to cool once under refrigeration; and
- length of refrigeration time until consumption.

The module can also be used to simulate the effect of intervention strategies. The inputs to the module are the regional and seasonal distributions of total and pathogenic *V. parahaemolyticus* at harvest. The output of the module is a series of predicted distributions of the total and pathogenic numbers at time of consumption. The final steps to be addressed in the exposure assessment are the storage and retail conditions of the product, storage after retail, and finally preparation and consumption. Some of these factors are included at the end of the flowchart presented in Figure I–6.

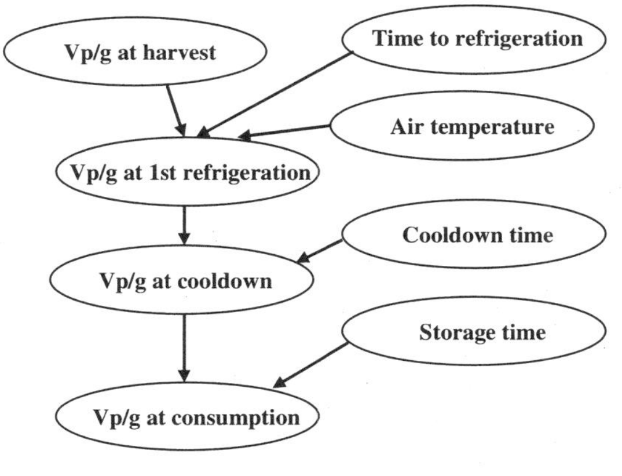

Figure I–6. Schematic depiction of the Post-harvest Module of the *V. parahaemolyticus* (Vp) risk assessment model.

Exposure assessment for Bloody clam

The production-to-consumption pathway for Bloody clam was used to estimate exposure (Figure II–6). This was based on data for total and pathogenic *V. parahaemolyticus* in 32 samples of Bloody clam studied at harvest and retail stages. The growth rate *k* was estimated based on the difference in counts and time from harvest to retail. This growth rate was used to estimate growth during subjectively estimated time between post-retail and consumption. All *V. parahaemolyticus* strains (pathogenic, *tdh*+ or *trh*+) were assumed to grow at the same rate. The fraction of Bloody clam meals that were improperly cooked was estimated by a small survey. It was assumed that all *V. parahaemolyticus* either survived improper cooking or were inactivated with proper cooking. Annual consumption was estimated based on survey data.

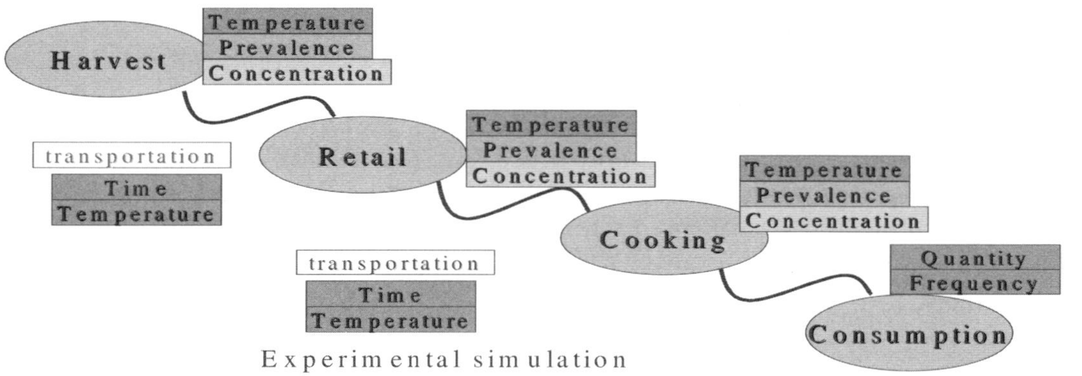

Figure II–6. Schematic representation of the model framework for a production-to-consumption risk assessment of *V. parahaemolyticus* in Bloody clam.

Exposure assessment for horse mackerel

Horse mackerel harvest and post-harvest practices in Japan formed the basis for developing the model for the production-to-consumption pathway. Eight different scenarios were modelled (Figure III–10). This included washing with clean water or no washing at the port; transportation in clean or contaminated water; washing or no washing of visceral cavity during the preparation; and various combinations of these. Data for prevalence and effect of washing on *V. parahaemolyticus* were available from earlier studies, and some studies were performed additionally for the risk assessment. The proportion of pathogenic *V. parahaemolyticus* was estimated based on random selection of data for other seafoods. Growth of *V. parahaemolyticus* during storage and transportation was estimated using the growth model from the FDA-VPRA risk assessment for raw oysters (FDA, 2005). Consumption data from various Japanese national surveys were used to estimate the meal size, proportion of horse mackerel consumed raw, and frequency of consumption.

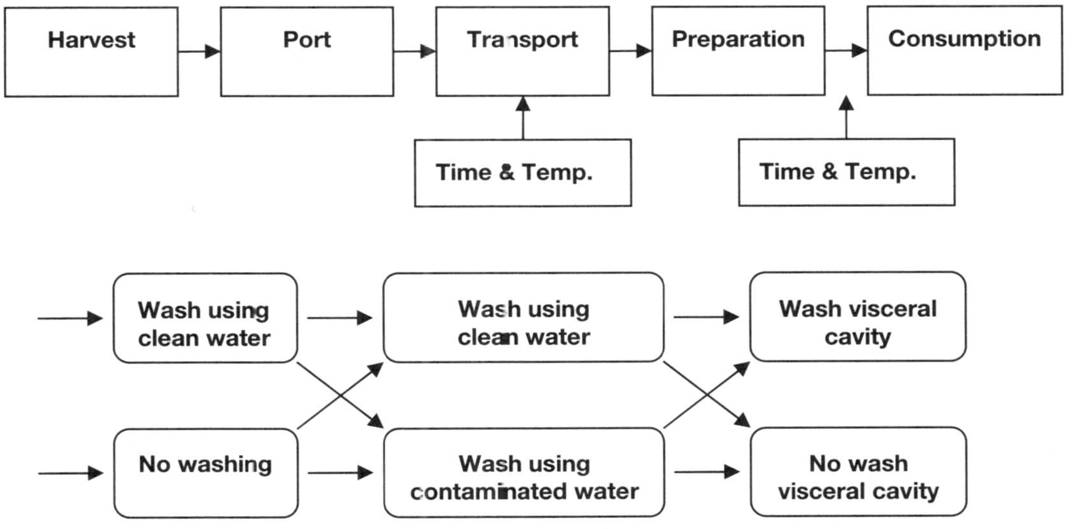

Figure III–10. Proposed scenarios at each stage and the eight scenarios modelled in the finfish (horse mackerel) risk assessment.

Hazard characterization

The dose response relationship between the amount of pathogenic *V. parahaemolyticus* consumed and the occurrence of illness was modelled as in the FDA-VPRA (FDA, 2005). To summarize: three human volunteer feeding studies were aggregated and a family of parameters for Beta Poisson curve fits were obtained using non-parametric bootstrapping (Figure I–9).

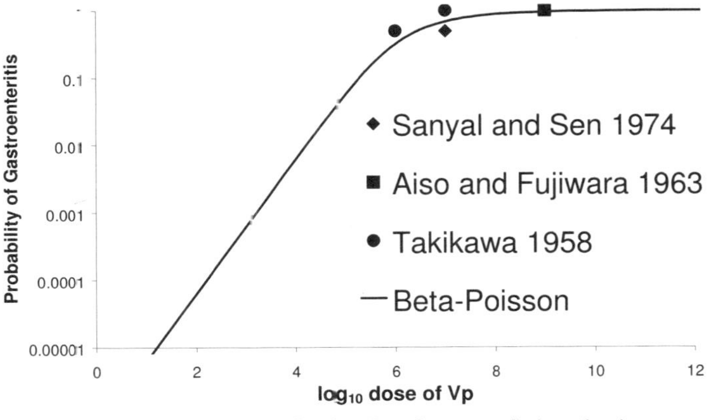

Figure I–9. Graphic representation of one of a family of curves fitting the human volunteer feeding studies used to characterize the risk associated with the consumption of pathogenic *V. parahaemolyticus*. For clarity, only one dose-response curve (the most likely) is shown.

Dose-response parameters were selected probabilistically during the simulations. For each simulation, the distribution of the risk of illness was recorded and the mean number of illnesses was obtained by multiplying the mean risk by the estimated number of oyster meals. Figure I–9 shows a representative plot of the dose response relationship.

Risk characterization

Raw oysters

Table I–27 presents estimates for predicted *V. parahaemolyticus* illness in the five countries modelled in the present risk assessment.

In Japan, production in Hiroshima Bay was taken as representative of that country's oyster production. The model predicts illnesses only in the October to December period as oysters are not harvested during the summer months.

In Australia, Wallis Lake was taken as typical of oyster harvesting areas of New South Wales, Australia, while Orongo Bay was considered representative of oyster production in New Zealand. Both areas were modelled for intertidal harvest.

Canadian production was modelled as for United States of America Pacific Northwest production, with changes in harvesting practice parameters. Canadian requirements are to bring oysters under temperature control within four hours of harvest when the air temperature is <15°C; at warmer temperatures, oysters are to commence cooling within one hour of harvest.

With the exception of New Zealand, the model predicted higher levels of illness in the growing area studied than are recorded in each country's notifications (Table I–28) where estimates are presented for predicted *V. parahaemolyticus* illness in the five countries modelled in the present risk assessment.

Table I–27. Predicted annual illnesses due to *V. parahaemolyticus* following consumption of oysters in Japan, Australia, New Zealand, Canada and United States of America.

	Jan–Mar	Apr–Jun	Jul–Sep	Oct–Dec
Hiroshima Bay, Japan	0	0	0	38
Wallis Lake, Australia	19	1	0	7
Orongo Bay, New Zealand	0	0	0	0
British Columbia, Canada	0	18	168	0
USA Gulf	10	698	1705	183
USA North Atlantic	0	3	14	2
USA Mid-Atlantic	0	4	7	4
USA Pacific Northwest	0	18	177	1

NOTE: The warmer months in the southern hemisphere are the opposite of those in the northern.

Table I–28. Predicted and reported annual illnesses due to *V. parahaemolyticus* following consumption of oysters in Japan, Australia, New Zealand and Canada.

	Annual cases predicted by model for the specified growing area	**Proportion (%) of total production**	**Annual cases predicted by model for nation**	**Epidemiological evidence for *V. parahaemolyticus* illness from oyster consumption**
Wallis Lake, Australia	27	30	91	2 cases in 18 years; two large outbreaks from other seafood sources
Orongo Bay, New Zealand	0	15	0	None during 1997–2002 from oysters; several outbreaks from other seafood sources
Hiroshima Bay, Japan	38	57.2	66	13 during 1998–2004
British Columbia, Canada	186	100	186	212 in decade 1997–2006

NOTE: The warmer months in the southern hemisphere are the opposite of those in the northern.

Models that predict illnesses from diseases with symptoms that range as a continuum from mild to severe provide special challenges in reporting. Ideally, the total number of illnesses estimated (with uncertainties) will be comparable with epidemiological surveillance data. However, for illnesses that present a range of symptoms, estimates will always be higher than reported numbers. The rate of under-reporting associated with the illness then becomes of pivotal importance.

In the present work, the CDC's adjustment of 20 unreported *V. parahaemolyticus* illnesses for every reported illness was used. Unfortunately, there is no information on the scale on which *V. parahaemolyticus* illness is unreported in Australia, Japan or New Zealand. What can be said is that each of these countries keeps statistics on illnesses caused by *V. parahaemolyticus*. In Japan, in particular, this pathogen is a major cause of food poisoning, though outbreaks and cases post-1999 seem to have diminished, perhaps due to the practice of harvesting oysters for raw consumption only when water temperatures are low. Cases reported for British Columbia are based predominantly on lab confirmation, with a minority of cases based on an association of clinical symptoms and food history. In Australia, there were large outbreaks of *V. parahaemolyticus* in New South Wales, involving chilled, cooked prawns imported from Indonesia. In 1990, more than 100 people were affected and one died; in 1992, two separate outbreaks involving >50 people occurred following consumption of cooked prawns received from the same wholesaler on the same day (Kraa, 1995). Since these outbreaks, *V. parahaemolyticus* has been included in routine laboratory screening of cases of suspected food poisoning in New South Wales.

The foregoing leads to the conclusion that under-reporting of *V. parahaemolyticus* illness in Japan, Canada, New Zealand and Australia is probably at least as extensive as in United States of America. Applying this ratio to Japan and British Columbia brings predicted and reported cases to the same order of magnitude, which may be considered an acceptable test for the model. In Australia, in contrast, there have been two reported illness occurrences in 18 years, compared with the model's prediction of around 1700 cases over the same period. Because the study of Lewis et al. (2003)

provides only preliminary information on prevalence and concentration of pathogenic *V. parahaemolyticus* from Wallis Lake oysters, data from the United States of America Pacific Northwest were used. These data, together with the water temperatures usual at Wallis Lake, account for the predicted illnesses. However, the study of Eyles, Davey and Arnold (1985) shows clearly that *V. parahaemolyticus* does not grow in Sydney rock oyster stored at 30°C for seven days.

Within the constraints imposed by lack of local data and knowledge of under-reporting, the present assessment provides estimates of illness of *V. parahaemolyticus* from oysters that are of the same order of magnitude for three countries and, while in the case of Australia the predicted versus reported case gap is greater, it may be that the oyster species studied is unique in its ability to prevent growth of *V. parahaemolyticus*.

Bloody clam

The upper estimate of the number of times a person would get ill with *V. parahaemolyticus* from consuming Bloody clam was 3.56×10^{-3} or a probability of 4 in 1000 persons per year. Epidemiological data shows that in Thailand in 1999 there were 319 isolations of *V. parahaemolyticus* in two hospitals in a city with a population of 282 000. This would be about 1 case per 1000, although the isolations would represent cases from all sources and the number of cases attributable to Bloody clam unknown. As discussed elsewhere, there would be under-reporting of cases and, in mild cases, stools might not be cultured. Taking the CDC ratio of 20 unreported cases for every reported case, even in Thailand, the estimate would be 20 cases per 1000 people, of which 4 could be due to Bloody clam. Thus it can be considered that the simulation model provides an estimate that is very close to epidemiological findings.

Horse mackerel

The risk assessment model estimated that the probability of becoming ill per serving of raw horse mackerel was 8.77×10^{-7} (best scenario) to 3.75×10^{-5} (worst scenario). For the Japanese population, the estimate was 70–1300 cases per year. Epidemiological data indicates around 10 000 cases annually, but the implicated seafoods are not always identified. The model was useful to estimate the impact of post-harvest practices on the probability of illness. Washing the eviscerated cavity of horse mackerel during preparation had a significant effect in reducing illness, while using disinfected water at ports and during transportation was much less effective (Table III–14).

Table III–4. Results of Monte Carlo simulation for selected scenarios: *V. parahaemolyticus* (Vp) numbers and probability of becoming ill per serving of raw horse mackerel.

Scenarios			Vp in whole body	Vp/g in fillet	Virulent Vp per serving	Probability of becoming ill per serving
Harvest	**Transport**	**Preparation**				
No washing	Contaminated water	No washing	6841	19.4	27.1	1.61E-05
		Washing		1.13	1.59	9.32E-07
	Clean water	No washing	6425	18.2	25.5	1.51E-05
		Washing		1.06	1.50	8.80E-07
Washing	Contaminated water	No washing	6225	17.7	24.7	1.47E-05
		Washing		1.03	1.45	8.48E-07
	Clean water	No washing	6134	17.4	24.3	1.44E-05
		Washing		1.02	1.43	8.40E-07

Impact of establishing limits for the level of V. parahaemolyticus *in oysters*

At the 38th session of the Codex Committee on Food Hygiene (CCFH) the Committee reviewed the hygiene provisions in the Proposed Draft Standard for Live and Raw Bivalve Molluscs, which had been developed by the Codex Committee on Fish and Fishery Products. It was concluded that, while the standard includes various microbiological limits for bivalves, the basis for the proposed limits was not clearly established and there was no agreement as to what these limits should be. The report of the session indicated that the CCFH

> "..... request FAO and WHO to use the risk assessment on *Vibrio parahaemolyticus* in seafood, which they are developing, to provide scientific guidance to the Codex Committee on Fish and Fishery Products to follow up on the recommendations of the CCFH regarding the hygiene provisions in the Proposed Draft Standard for Live and Raw Bivalve Molluscs."

The risk management question posed was:

> "Estimate the risk reduction from *V. parahaemolyticus* when the total number of *V. parahaemolyticus* or the number of pathogenic *V. parahaemolyticus*, ranges from absence in 25 g to 1000 cfu or MPN per gram."

The estimation considered the impact of three different limits for *V. parahaemolyticus*: 100 cfu/g, 1000 cfu/g and 10 000 cfu/g. These limits were considered to be applied when the products are cooled after harvesting, when the population of *V. parahaemolyticus* has stabilized, i.e. when the temperature becomes too low for further growth but not so low that die-off occurs.

The estimation looked at the impact of all (100%) harvested oysters meeting a specified target limit compared with the baseline distribution of *V. parahaemolyticus* for each of these countries, and it estimated the reduction in human illness, together with the amount of product rejection, that would occur if all product on the market were to meet the specified target. The estimates are presented in Tables I–32 and I–33 and the caveats applying to the estimates are covered in detail in Section I–5.1 of the main report.

Table I–32. Reduction in illness, based on meeting specified target numbers of *V. parahaemolyticus*, together with commensurate rejection of product for raw consumption.

Target specified	Reduction (%) in the number of predicted illnesses			Product (%) rejected to achieve these reductions in illness		
	Australia (summer)	**New Zealand (summer)**	**Japan (autumn)**	**Australia (summer)**	**New Zealand (summer)**	**Japan (autumn)**
100 cfu/g	99	96	99	67	53	16
1000 cfu/ g	87	66	97	21	10	5
10 000 cfu/g	52	20	90	2	1	1

Table I–33. Predicted *V. parahaemolyticus* illnesses as a result of meeting specified targets.

	Predicted number of illnesses per year		
Specified target	**Australia (summer)**	**New Zealand (summer)**	**Japan (autumn)**
100 cfu/g	Approx 1 every 5 years	Approx 1 every 10 years	Approx 1 every 2 years
1000 cfu/ g	1	1	1
10 000 cfu/g	5	3	4
No limit	17	4	38

Using the risk assessment tool to inform industry process regimes

The models have utility to inform industry of the outcome of various "what-if" scenarios. An example of one such scenario is the enhanced chilling of oysters after harvest from waters of different temperature (waters in different seasons), developed by the US FDA as part of the *V. parahaemolyticus* Control Plan adopted by the General Assembly of the 2007 meeting of the Interstate Shellfish Sanitation Conference (ISSC). The scenario demonstrates the effectiveness of limiting ambient exposure prior to refrigeration, coupled with more rapid cooling, on reduction of *V. parahaemolyticus* illnesses from consumption of oysters harvested at various water temperatures. In the case of horse mackerel, washing the eviscerated cavity during preparation of *sashimi* would contribute to reduced illness.

Key outcomes of the risk assessment

Raw oyster risk assessment

The present risk assessment operated under constraints resulting from incomplete data, as summarized in Tables I-1 and I–26 in the main report. Surrogate data were required to enable the model to be run. Some data gaps went to the very heart of the risk assessment, such as the prevalence of *V. parahaemolyticus* in oysters, coupled with the proportion which were *tdh+*. Perhaps unsurprisingly, given the foregoing, estimates of illness for the four countries that were able to supply some required data were not in accord with surveillance data for each country.

However, the present work has generated some positive outcomes:

- The model has been used to estimate illness from different oyster species grown under various regimes and regulatory management systems.
- The framework of the model is made available in this report for modification by risk assessors in other countries.
- Risk assessors will benefit from the accumulated knowledge presented in this report.
- Some countries, e.g. New Zealand and Australia, are investing in significant research studies on *V. parahaemolyticus* prevalence in oysters, and the proportion of pathogenic strains.
- The model is an appropriate tool for testing mitigation strategies both at harvest (such as reduced cooling times) or post-harvest by heating, freezing or high-pressure treatment.
- The United States of America has used the FDA-VPRA to enhance industry practice and management.

Bloody clam risk assessment

This is a good example demonstrating that risk assessment models with data collected in a city for a limited period of time can be used to estimate the public health risk due to consumption of a popular seafood item at national level. Though the public perception (data from consumer survey) is that Bloody clam consumption is associated with incidents of diarrhoea in Thailand, the estimate from the model seems to be close to epidemiological finding from two hospitals in the city.

Horse mackerel risk assessment

Though the risk assessment was carried out under the constraints noted, this illustrates how the model derived from raw oysters can be extended to estimate risk through consumption of raw finfish. This assessment provides a good example of using the model for estimating risk reductions that can be achieved by post-harvest practices like washing the eviscerated cavity of horse mackerel during preparation of *sashimi*.

Gaps in the data

Deficiencies in the data available to conduct the present risk assessment were identified in order to suggest future research or further data gathering to reduce uncertainties. The areas requiring further study are:

- Incidence and frequency of pathogenic *V. parahaemolyticus* in water, finfish and shellfish.
- Factors that affect incidence of pathogenic *V. parahaemolyticus* in the environment.
- Role of oyster physiology and immune status in levels of *V. parahaemolyticus*. There is a need to correlate the number of *V. parahaemolyticus* with the percentage of oysters contaminated.
- Determining whether the dose response model currently available only for *tdh*+ strains could be different for pandemic strains and for *trh*+ strains.
- More research on the potential virulence factors of pathogenic strains other than TDH, e.g.TRH1 and TRH2 enterotoxins. *V. parahaemolyticus* strains that do not produce TDH, TRH or urease have recently been found to induce fluid accumulation in suckling mice and diarrhoea in a ferret model after oral inoculation in a dose-dependent manner (Kothary et al., 2000). Correlation between clinical and environmental incidence of these strains had yet to be determined.
- Growth rates of *V. parahaemolyticus* within different oyster species at temperatures other than 26°C, including the issue of potential differences in the growth rate of pathogenic strains versus total *V. parahaemolyticus* populations.
- Growth rates of *V. parahaemolyticus* in other shellfish and finfish under different post-harvest handling and storage conditions.
- Rates of hydraulic flushing (water turnover) in shellfish harvest areas based on levels of freshwater flows, tidal changes, winds and depth of harvesting area, and how these factors may influence pathogenic *V. parahaemolyticus* numbers.
- Consumer handling of oysters.
- Cross-contamination when finfish are prepared for raw consumption, e.g. for *sashimi* and *sushi*.
- Improved global public health surveillance of *V. parahaemolyticus* to identify new epidemic strains as they emerge.
- Knowledge of reporting systems in each country of study; this is of specific importance when under-reporting is taken into account to compare estimated with reported illnesses.

Current and future issues influencing risk assessment and management of V. parahaemolyticus in oysters

Risk assessments are typically conducted over a significant period and the oyster assessment is no exception, having its genesis in 2001. It is axiomatic that, given the pace of scientific progress, any risk assessment process will always lag somewhat. In an attempt to accommodate recent advances and events surrounding *V. parahaemolyticus*, the present section attempts to provide an up-to-date scientific context for the key premises underpinning the assessment, and speculates on how this risk assessment model might be used and developed by risk managers as growing waters become progressively warmer and *V. parahaemolyticus* extends its latitudinal reach.

This risk assessment predicts the risk of *V. parahaemolyticus* illness from consumption of raw oysters harvested from selected Pacific Rim countries based on the framework previously developed to predict risk of illness in the United States of America (FDA, 2005). It was constructed around four key premises:

- The risk of infection is proportional to the exposure to pathogenic strains, defined as those possessing the *tdh* gene, and that all strains possessing *tdh* are equally pathogenic.
- Levels of pathogenic *V. parahaemolyticus* can be reliably estimated from total *V. parahaemolyticus* levels, which are easier to measure.
- *V. parahaemolyticus* levels in oysters at harvest can be predicted based on water temperature.
- Total and pathogenic *V. parahaemolyticus* grow and survive equally during post-harvest handling and processing.

In the FDA-VPRA (FDA, 2005), a market study of raw oysters that was not used to estimate model parameters was used to validate the exposure assessment. The reliability of the VPRA's ability to predict exposure to total *V. parahaemolyticus* levels at or near the point of consumption was demonstrated for each season. However, regional attribution of illnesses as determined by the risk characterization was not consistent with estimates based solely on reported illnesses for which trace-back information was available. While some of these differences may be due to regional biases in reporting of illness and/or trace-back, they also suggest that some premises, or assumptions, underpinning the VPRA may be suspect. Considerable new information on these key premises has become available since the FDA and FAO/WHO VPRA's were initiated. These new findings are discussed below and may provide guidance for future efforts to refine the current versions of the VPRA and to implement these models for risk management.

The premise that *V. parahaemolyticus* risk is proportional to exposure to levels of pathogenic *V. parahaemolyticus* continues to be supported by epidemiological data, as ca. 90% of clinical isolates reported to CDC in the 2001–2004 period possessed the *tdh* gene. However, an increasing proportion of clinical isolates possessed neither the *tdh* nor the *trh* gene, and these were associated with the most severe cases, requiring hospitalization. It is unknown whether these isolates are related to each other, nor the types of virulence attributes they possess. One of these strains has been selected for whole-genome sequencing; mining of the resultant data may reveal potential virulence determinants. The assumption that all *tdh*+ strains are equally virulent was based more on a lack of data to the contrary rather than targeted studies testing this hypothesis. Epidemiological studies of the 2004 Alaskan cruise ship outbreak have produced perhaps the strongest challenge to this assumption. The outbreak was associated with a strain that was serotype O6:K18, which caused a 30% attack rate among passengers consuming one to six oysters during three cruises over a two-week period in July 2004 (McLaughlin et al., 2005). Oysters collected from the implicated farm during this period typically contained less than 10 MPN per gram of the outbreak strain. The VPRA dose response curve indicates that a dose ca. 10 000-fold greater would be necessary to cause a 30% attack rate. The O6:K18 isolates from the Alaskan outbreak were indistinguishable by PFGE from those isolated in sporadic cases from Pacific Coast States over the previous decade. It is unclear whether this strain can infect at lower doses than the strains used in previous human volunteer studies. Another possible explanation is that the liquid doses used in human volunteer studies may underestimate the infectivity of *V. parahaemolyticus* relative to consumption of raw oysters that have naturally accumulated *V. parahaemolyticus*.

New studies employing real time PCR analysis of oyster alkaline peptone water enrichments in an MPN format are addressing the second key premise of this VPRA, namely that levels of pathogenic *V. parahaemolyticus* can be reliably estimated from total *V. parahaemolyticus* levels (Miwa et al., 2003; Nordstrom et al., 2007). This PCR-MPN approach permits simultaneous examination of thousands of *V. parahaemolyticus* cells for *tdh* and *trh* genes from each MPN tube and is much more sensitive than cultural methods that require colony isolation and typically only examine a few isolates for these genes. However, the recent detection of *trh* gene in other *Vibrio* spp., such as *V. alginolyticus* and *V. harveyi* (Gonzalez-Escalona et al., 2006; Masini et al., 2007) complicates the interpretation of such assays. Even among *tdh*+ strains, expression of the gene varies depending on the type of allele present. Expression of TDH is much higher in strains possessing the *tdh*2 gene compared with other variants of *tdh* (Nishibuchi and Kaper, 1995; Nakaguchi and Nishibuchi, 2005). Thus strains with other variants of the *tdh* gene may be less pathogenic compared with strains possessing the *tdh*2 gene. These studies are beginning to show that the relationship between total and pathogenic *V. parahaemolyticus* is far more variable than assumed in the VPRA. For instance, ecological studies of *V. parahaemolyticus* in Alaska indicated that pathogenic *V. parahaemolyticus* accounted for 74% and 30% of the total *V. parahaemolyticus* population of oysters in the summers of 2004 and 2005, respectively. Employing the same methods in Alabama and Mississippi in the summer of 2004, pathogenic *V. parahaemolyticus* accounted for 0.001% and 0.05% of the *V. parahaemolyticus* population, respectively (DePaola et al., 2007). Thus consumption of Alaskan oysters containing only 10 *V. parahaemolyticus* per gram may be more risky than consuming Alabama oysters with levels greater than 100 000

V. parahaemolyticus per gram, assuming that the *tdh*+ strains from Alaska and Alabama are equally virulent.

The third key premise of the VPRA, that *V. parahaemolyticus* levels in oysters at harvest can be predicted by water temperature, still appears to be on fairly solid ground for total *V. parahaemolyticus*. When water temperatures are below 15°C, *V. parahaemolyticus* levels are generally below one per gram and outbreaks do not occur. However, data is beginning to accumulate indicating that the ratio of pathogenic to total *V. parahaemolyticus* may be affected by water temperature, with higher prevalence of pathogenic *V. parahaemolyticus* occurring at lower water temperatures. Regionally, the proportion of *V. parahaemolyticus* populations in United States of America oysters possessing the *tdh* gene appears to be increasingly moving north from the Gulf Coast towards the Pacific Northwest and Alaska. In a study of Alabama oysters, the proportion of pathogenic *V. parahaemolyticus* was found to be ca. 10-fold greater when the water temperature was <20°C compared with >20°C (De Paola et al., 2003). Water temperature alone has been estimated to account for approximately 50% of the annual variability in total *V. parahaemolyticus* levels of oysters. Recent studies suggest that salinity and turbidity may also influence *V. parahaemolyticus* levels in oysters, and incorporation of these parameters into the *V. parahaemolyticus* model may further reduce uncertainty in model predictions. Water temperature, salinity and turbidity can all be measured remotely by satellite imagery on a daily basis worldwide. The feasibility of integrating risk assessment models into remote sensing data for prediction of *V. parahaemolyticus* levels in real time has been demonstrated in a recent scientific publication. This approach was suggested as a management tool for *V. vulnificus* in oysters at the 2006 FAO/WHO Kiel Conference for developing practical risk management applications from quantitative microbiological risk assessments (FAO/WHO, 2006).

There has been limited research addressing the fourth and final key premise, namely that total and pathogenic *V. parahaemolyticus* grow and survive equally during post-harvest handling and processing. In terms of controlling oyster-borne *V. parahaemolyticus* illnesses, this is probably the most important premise, since harvest is the point when humans take over control of *V. parahaemolyticus* levels from nature. FDA submitted a proposal to the 2007 ISSC in Albuquerque, United States of America, to limit *V. parahaemolyticus* growth in oysters post-harvest to half a log. The VPRA predicts that achieving an average of a half-log reduction in *V. parahaemolyticus* levels in oysters will result in ca. 70% reduction in *V. parahaemolyticus* illnesses associated with consumption of United States of America oysters. Preliminary results from storage of Alaskan oysters at ambient air temperature (15–20°C) for up to 5 days indicated that growth rates of total and pathogenic *V. parahaemolyticus* were nearly identical and were in agreement with growth rates predicted by the VPRA at these temperatures. There is some information that the O3:K6 pandemic strains are more pressure resistant than other *V. parahaemolyticus* strains (Cook, 2003). However, in an FDA/ISSC survey of post-harvest treated oysters in the United States of America (mild pasteurization, freezing or high hydrostatic pressure), approximately 40 *V. parahaemolyticus* strains that survived these treatments were isolated, primarily from 25-gram enrichments (unpublished data). A single *tdh*+

isolate was observed among these strains, suggesting that *tdh+ V. parahaemolyticus* responds to these post-harvest treatments in a similar manner to non-pathogenic *V. parahaemolyticus*.

In the past decade, trends in *V. parahaemolyticus* illnesses associated with molluscan shellfish have taken unexpected and unprecedented turns. Pandemic *V. parahaemolyticus* has now spread to at least 5 continents, and has caused repeated annual outbreaks in Chile, where *V. parahaemolyticus* was not even considered a problem when this risk assessment began in 2001. One of the outbreaks in Chile was of epidemic proportions, causing over 10 000 cases, most of which were associated with consumption of molluscan shellfish (Fuenzalida et al., 2006). More and more of the scientific community are coming to the conclusion that ballast discharge is the major mechanism for global spread of pandemic *V. parahaemolyticus*. While global trade continues to show unprecedented growth, there are few if any control measures implemented to protect molluscan shellfish from contamination. Vibrios appear to be uniquely adapted in the microbe world for global dissemination by ballast discharge, and, in the absence of controls, it is likely that new strains will emerge and be spread globally.

In the past decade, the warmest temperatures on record have occurred in many parts of the world that are major producers of molluscan shellfish, and even warmer conditions are forecast for coming decades. Warmer temperatures may in part be responsible for the CDC reported increase of *Vibrio* infections in United States of America, while illnesses due to other major foodborne pathogens are declining (CDC, 2007). Since 2000, November is the leading month for *V. vulnificus* infections associated with oyster consumption in the United States of America (Mark Glatzer, FDA, pers. comm.). The warmer temperatures also appear to be the cause of *V. parahaemolyticus* extending its geographical range into areas such as Alaska (McLaughlin et al., 2005), Europe (Martinez-Urtaza et al., 2005) and Chile (Gonzales-Escalona et al., 2005). Alaska is 1000 km further north than any previously reported *V. parahaemolyticus* illnesses. The unprecedented high prevalence of pathogenic strains among the *V. parahaemolyticus* population of Alaskan oysters, and perhaps higher virulence compared with other pathogenic strains, was especially surprising and remains unexplained.

The Alaskan oyster industry could not wait for elucidation of *V. parahaemolyticus* ecology in the Prince William Sound, as their industry was fragile and at risk of collapsing if outbreaks continued. Instead they relied heavily on the FDA VPRA, and in 2005 they voluntarily adopted the pro-active control of sinking oyster cages below the thermocline where water temperatures were below 10°C. Under similar environmental conditions as in 2004, when 62 *V. parahaemolyticus* cases were reported, there was only a single case reported in 2005 after implementing this control (Manny Soares, Alaska Department of Conservation, pers. comm.). In response to the 1997 and 1998 *V. parahaemolyticus* outbreaks in the United States of America Pacific Northwest, the Canadian government implemented strict time × temperature controls for oysters harvested in British Columbia, and no major outbreaks have since occurred. Washington State, which has similar environmental conditions to British Columbia, but less stringent time × temperature controls, experienced a major outbreak in 2006 of 177 cases, most of which were associated with consumption of raw oysters (CDC, 2006). In most areas of the world it will probably take decades, if ever, before a system is available to reliably predict levels of pathogenic *V. parahaemolyticus* in the environment. As new and more powerful methods for measuring pathogenic *V. parahaemolyticus*, such as real time PCR, obtain more widespread acceptance and application, areas and times of high risk should become more apparent. At the same time, the effectiveness of pro-active post-harvest controls based on the existing VPRA have already been demonstrated in the real world, and this approach currently offers the best available opportunity to control *V. parahaemolyticus* illnesses from consumption of raw oysters and other raw molluscan shellfish.

References for Interpretative Summary

Aiso, K. & Fujiwara, K. 1963. Feeding tests of the pathogenic halophilic bacteria. *Annals of the Research Institute for Food Microbiology of Chiba University*, 15: 34–38.

Anon[ymous]. 1998a. *Bol. Epidemiol. Semenal*, 6(31): 306.

Anon. 1998b. National Nutrition Survey, Japan. Ministry of Health and Welfare, Japan.

Anon. 1999a. *Vibrio parahaemolyticus*, Japan 1996–1998. *Infectious Agents Surveillance Report*, 20(7): 1–2.

Anon. 1999b. *Vibrio parahaemolyticus*, Taiwan: Background. *PROMED-digest*, 28 May 1999.

CDC [Centres for Disease Control]. 2006. *Vibrio parahaemolyticus* infections associated with consumption of raw shellfish – three states, 2006. *Morbidity and Mortality Weekly Report*, 55: 854–856.

CDC. 2007. Preliminary FoodNet data on the incidence of infection with pathogens transmitted commonly through food – ten states, 2006. *Morbidity and Mortality Weekly Report*, 56: 336–339.

Cook, D.W. 2003. Sensitivity of *Vibrio* species in phosphate-buffered saline and in oysters to high pressure processing. *Journal of Food Protection*, 66(12): 2276–2282.

DePaola, A., Nordstrom, J.L., Bowers, J.C., Wells, J.G. & Cook, D.W. 2003. Seasonal abundance of total and pathogenic *Vibrio parahaemolyticus* in Alabama oysters. *Applied and Environmental Microbiology*, 69: 1521–1526.

DePaola, A., Nordstrom, J.L., Blackstone, G.M., Bowers, J.C., Phillips, A.M.B., Grimes, D.J., Watt, S. & Parveen, S. 2007. Enumeration of total and pathogenic *Vibrio parahaemolyticus* using real-time PCR in a most probable number format. p. 36, in: [Proceedings of the] 6th Annual International Conference on Molluscan Shellfish Safety, Blenheim, New Zealand.

Eyles, M., Davey, G. & Arnold, G. 1985. Behaviour and incidence of *Vibrio parahaemolyticus* in Sydney rock oysters (*Crassostrea commercialis*). *International Journal of Food Microbiology*, 1: 327–334.

European Commission. 2001. Opinion of the Scientific Committee on Veterinary Measures relating to Public Health on *Vibrio vulnificus* and *Vibrio parahaemolyticus* in raw and undercooked seafood. Available at http://ec.europa.eu/food/fs/sc/scv/out45_en.pdf

FAO. 2005. Causes of detentions and rejections in international trade. Prepared by L. Ababouch, G. Gandini and J. Ryder. *FAO Fisheries Technical Paper* No. T473. 110 p.

FAO/WHO. 2006. Report of the Joint FAO/WHO Expert Consultation on the use of microbiological risk assessment outputs to develop practical risk management strategies: Metrics to improve food safety. Kiel, Germany, 3–7 April 2006. Available at: ftp://ftp.fao.org/ag/agn/food/kiel.pdf

FDA [US Food and Drug Administration]. 2005. Quantitative Risk Assessment on the Public Health Impact of Pathogenic *Vibrio parahaemolyticus* in Raw Oysters. U.S. Department of Health and Human Services, U.S. Food and Drug Administration. See: http://www.fda.gov/Food/ScienceResearch/ResearchAreas/RiskAssessmentSafetyAssessment/ucm185746.htm

Fuenzalida, L., Hernandez, C., Toro, J., Rioseco, M.L., Romero, J. & Espejo, R.T. 2006. *Vibrio parahaemolyticus* in shellfish and clinical samples during two large epidemics of diarrhoea in southern Chile. *Environmental Microbiology*, 8(4): 675–683.

Fujino, T., Okuno, Y., Nakada, D., Aoyoma, A., Fukai, K., Mukai, T. & Ueho, T. 1953. On the bacteriological examination of shirasu food poisoning. *Medical Journal of Osaka University*, 4: 299–304.

Gonzalez-Escalona, N., Cachicas, V., Acevedo, C., Rioseco, M.L., Vergara, J.A., Cabello, F., Romero, J. & Espejo, R.T. 2005. *Vibrio parahaemolyticus* diarrhea, Chile, 1998 and 2004. *Emerging Infectious Diseases*, 11: 129–131.

Hlady, W.G. 1997. *Vibrio* infections associated with raw oyster consumption in Florida, 1981–1994. *Journal of Food Protection*, 60: 353–357.

Honda, T., Ni, Y.X. & Miwatani, T. 1988. Purification and characterization of a hemolysin produced by a clinical isolate of Kanagawa phenomenon-negative *Vibrio parahaemolyticus* and related to the thermostable direct hemolysin. *Infection and Immunity*, 56(4): 961–965.

Kothary, M.H., Burr, D.H., Tall, B.D., Hanes, D.E. & Miliotis, M.D. 2000. Thermostable direct hemolysin, thermostable-related hemolysin, and urease are not required for pathogenicity of *Vibrio parahaemolyticus* in animal models. p. 68, in: Abstracts of the 100th General Meeting of the American Society for Microbiology.

Kraa, E. 1995. Surveillance and epidemiology of foodborne illness in NSW, Australia. *Food Australia*, 4: 418–423.

Lewis, T., Brown, M., Abell, G. et al., 2003. Pathogenic *Vibrio parahaemolyticus* in Australian oysters. Fisheries Research and Development Corporation, Report 2002/49.

Madigan, T., Lee, K., Pointon, A. & Thomas, C. 2007. A supply chain assessment of marine vibrios in Pacific oysters in South Australia: prevalence, quantification and public health risk. Fisheries Research and Development Corporation, Project SIDF 2005/401.

Martinez-Urtaza, J., Simental, L., Velasco, D., DePaola, A., Ishibashi, M., Nakaguchi, Y., Nishibuchi, M.,Carrera-Flores, D., Rey-Alvarez, C. & Pousa, A. 2005. Pandemic *Vibrio parahaemolyticus* O3:K6, Europe. *Emerging Infectious Diseases*, 11: 1319–1320.

Masini, L., De Grandis, G., Principi, F., Mengarelli, C. & Ottaviani, D. 2007. Research and characterisation of pathogenic vibrios from bathing waters along the Conero Riviera (Central Italy). *Water Research*, 41(18): 4031–4040.

McLaughlin, J.B., DePaola, A., Bopp, C.A., Martinek, K.A., Napolilli, N.P., Allison, C.G., Murray, S.L., Thompson, E.C., Bird, M.M. & Middaugh, J.P. 2005. Outbreak of *Vibrio parahaemolyticus* gastroenteritis associated with Alaskan oysters. *New England Journal of Medicine*, 353: 1463–1470.

Mead, P.S., Slutsker, L., Dietz, V., McCaig, L.F., Bresee, J.S., Shapiro, C., Griffin, P.M. & Tauxe, R.V. 1999. Food-related illness and death in the United States. *Emerging Infectious Diseases*, 5: 607–625.

Miles, D.W., Ross, T., Olley, J. & McMeekin, T.A. 1997. Development and evaluation of a predictive model for the effect of temperature and water activity on the growth rate of *Vibrio parahaemolyticus*. *International Journal of Food Microbiology*, 38: 133–142.

Miwa, N., Nishio, T., Kawamori, F., Masuda, T. & Akiyama, M. 2003. Evaluation of MPN method combined with PCR procedure for detection and enumeration of *Vibrio parahaemolyticus* in seafood. *Shokuhin Eiseigaku Zasshi*, 44: 289–293.

Nair, G.B., Ramamurthy, T., Bhattacharya, S.K., Datta, B., Takeda, Y. & Sack, D.A. 2007. Global dissemination of *Vibrio parahaemolyticus* O3:K6 and its serovariants. *Clinical Microbiological Reviews*, 20: 39–48.

Nakaguchi, Y. & Nishibuchi, M. 2005. The promotor region rather than its downstream inverted repeat sequence is responsible for low-level transcription of the thermostable direct hemolysin-related hemolysin (*trh*) gene of *Vibrio parahaemolyticus*. *Journal of Bacteriology*, 187: 1849–1855.

Nishibuchi, M. & Kaper, J.B. 1995. Thermostable direct hemolysin gene of *Vibrio parahaemolyticus*: a virulence gene acquired by this bacterium. *Infection and Immunology*, 63: 2093–2099.

Nordstrom, J.L., Vickery, M.C.L., Blackstone, G.M., Murray, S.L. & DePaola, A. 2007. Development of a multiplex real-time PCR assay with an internal amplification control for the detection of total and pathogenic *Vibrio parahaemolyticus* in oysters. *Applied and Environmental Microbiology*, 73(18): 5840–5847.

Pan, T.M., Chai, T.-J., Lee, C.L., Chien, S.W. & Horng, C.B. 1997. Foodborne disease outbreaks due to bacteria in Taiwan, 1986 to 1995. *Journal of Clinical Microbiology*, 35(5): 1260–1262.

Parveen, S., Tamplin, M.L., da Silva, L.V.A., White, C., Bowers, J.C., Rutto, G. & DePaola, A. 2007. Predictive models for the growth and survival of total *Vibrio parahaemolyticus* in Gulf Coast shellstock oysters. pp. 176–177, in: Abstracts of the 94th Annual Meeting of the International Association for Food Protection, Lake Bueno Vista, Florida, United States of America.

Sakazaki, R., Iwanami, S. & Tamura, K. 1968. Studies on enteropathogenic facultatively halophilic bacterium, *Vibrio parahaemolyticus*. II Serological characteristics. *Japanese Journal of Medical Science and Biology*, 21: 313–324.

Sanyal, S.C. & Sen, P.C. 1974. Human volunteer study on the pathogenicity of *Vibrio parahaemolyticus*. pp. 227–230, in: T. Fujino, G. Sakaguchi, R. Sakazaki and Y. Takeda. (editors). International Symposium on *Vibrio parahaemolyticus*. Saikon Publishing Company, Tokyo.

Takikawa, I. 1958. Studies on pathogenic halophilic bacteria. *Yokohama Medical Bulletin*, 9: 313–322.

TECHNICAL REPORT

Part I

Microbiological risk assessment of *Vibrio parahaemolyticus* in raw oysters

Introduction and scope

Statement of purpose

The purpose of the present assessment is to use the model developed during the United States of America Quantitative Risk Assessment on the Public Health Impact of Pathogenic *Vibrio parahaemolyticus* in Raw Oysters (FDA-VPRA) (FDA, 2005) to estimate risk of illness in other countries as a result of consumption of oysters.

Constraints

When the present study was initiated, it was hoped that country-specific data would be supplied by a number of countries. In the event, only Australia, Canada, Japan and New Zealand were able to participate, and were able to supply only some of the data needed to run the model. Where country-specific data were lacking, United States of America surrogate data were used, placing a major constraint on the present project. The scientific underpinning for the present study was published as FDA (2005)[2].

I–1. Hazard identification

Vibrio parahaemolyticus is a marine micro-organism native in estuarine waters throughout the world. The organism was first identified as a foodborne pathogen in Japan in the 1950s (Fujino et al., 1953). By the late 1960s and early 1970s, *V. parahaemolyticus* was recognized as a cause of diarrhoeal disease worldwide, although most common in Asia and the United States of America. A recent history of seafood consumption is a consistent aspect of *V. parahaemolyticus* infection. Vibrios concentrate in the gut of filter-feeding molluscan shellfish, such as oysters, clams and mussels, where they multiply and cohere. Although thorough cooking destroys these organisms, oysters are often eaten raw and, at least in the United States of America, are the most common food associated with *V. parahaemolyticus* infection (Hlady, 1997). Early studies in Japan showed that 96% of clinical strains produce a thermostable direct haemolysin (TDH), while only 1% of the environmental strains produce this haemolysin (Sakazaki et al., 1968). Subsequently, TDH-negative strains from clinical cases were found to produce a TDH-related haemolysin, TRH (Honda et al., 1988). Currently, strains producing TDH and TRH are considered pathogenic to man. Diverse serotypes may be associated with human infections, but, recently, strains belonging to the O3:K6 serotype have been found to be the causative agent of several outbreaks in different countries (Nair et al., 2007).

2. See http://www.fda.gov/Food/ScienceResearch/ResearchAreas/RiskAssessmentSafetyAssessment/ucm050421.htm

I–1.1 Human incidence

In Asia, *V. parahaemolyticus* is a common cause of foodborne disease. In general, the outbreaks are small in scale, involving fewer than 10 cases, but occur frequently. Prior to 1994, the incidence of *V. parahaemolyticus* infections in Japan had been declining; however, in 1994–95 there were 1280 reports of infection due to the organism (Anon., 1999a), and *V. parahaemolyticus* food poisonings outnumbered those of *Salmonella* food poisoning during this period. From 1996 to 1998, there were 496 outbreaks and 24 373 cases of *V. parahaemolyticus* reported, while from 1999 to 2005, 25 211 cases were reported (data from Ministry of Health, Labour and Welfare, Japan). In general, outbreaks were more prevalent in the summer, with a peak in August. Boiled crabs caused one large-scale outbreak, involving 691 cases. The increased incidence during 1997–1998 has been attributed to an increased incidence of serovar O3:K6.

Between 1986 and 1995, some 197 outbreaks of foodborne disease were caused by *V. parahaemolyticus* in Taiwan (Pan et al., 1997), while in 1997 over 200 outbreaks were reported, including an outbreak of 146 cases acquired from boxed lunches (Anon., 1999b).

During 1997 and 1998, there were more than 700 cases of illness due to *V. parahaemolyticus* in the United States of America, the majority of which were associated with the consumption of raw oysters. In two of the 1998 outbreaks a serotype of *V. parahaemolyticus,* O3:K6, reported previously only in Asia, emerged as a principal cause of illness for the first time. Subsequent studies on these strains have revealed their pandemic spread. It was suggested that warmer than usual water temperatures were responsible for the outbreaks

In Europe, few data exist on the incidence of *V. parahaemolyticus* infections, one reason being that such infections are not notifiable. However, the current knowledge of the incidence in Europe has been presented in *Opinion of the Scientific Committee on Veterinary Measures relating to Public Health on* Vibrio vulnificus *and* Vibrio parahaemolyticus *in raw and undercooked seafood* (European Commission, 2001). The data are summarized in Table I–1.

I–1.2 Foods implicated

V. parahaemolyticus occurs in a variety of fish and shellfish, including clams, shrimp, lobster, crayfish, scallops and crabs, as well as in oysters. Although oysters are the most common food associated with *Vibrio* infection in some countries (Hlady, 1997), there have been reports of *V. parahaemolyticus* infections associated with the other types of seafood. One such report was a case-control study of sporadic *Vibrio* infections in two coastal areas of Louisiana and Texas in the United States of America, conducted in 1992-93, in which crayfish consumption was reported by 50% (5/10) of the persons affected with *V. parahaemolyticus* infection (Bean et al., 1998).

Outbreaks of *V. parahaemolyticus* gastroenteritis aboard two Caribbean cruise ships were reported in 1974 and 1975 (Lawrence et al., 1979). The outbreaks were most likely caused by contamination of cooked seafood by seawater from the ships' seawater fire systems.

In 1972, an estimated 50% of 1200 persons who attended a shrimp feast in Louisiana in the United States of America became ill with *V. parahaemolyticus* gastroenteritis (Barker and Gangarosa, 1974), and samples of uncooked shrimp tested positive for the organism.

Three outbreaks occurred in Maryland in the United States of America in 1971 (Dadisman et al., 1972). Steamed crabs were implicated in two of the outbreaks after cross-contamination with live crabs. The third outbreak was associated with crabmeat that had become contaminated before and during canning.

Recently, sampling studies in the Adriatic Sea demonstrated the presence of *V. parahaemolyticus* in fish, mussels and clams (Baffone et al., 2000), and in a recent study of mussels from the northwestern coast of Spain, *V. parahaemolyticus* was isolated from 8% of samples (European Commission, 2001).

Table I–1. Available data on the incidence of *V. parahaemolyticus* infections in Europe.

Country	Period	No. of cases	Symptoms	Origin of data
Denmark	1987-1992	13 10	Wound infection Ear infection	Hornstrup and Gahrn-Hansen, 1993
Denmark	1980-2000	2	Gastroenteritis	Statens Serum Institut, Copenhagen, Denmark
England and Wales, UK	1995-1999	115		PHLS, Colindale, United Kingdom
France	1995-1998	6 1	Gastroenteritis Septicaemia	Geneste et al., 2000
France	1997	44	Gastroenteritis (1)	Lemoine et al., 1999
Northern Ireland, UK	1990-1999	0		CDSC (Communicable Disease Surveillance Centre, NI, United Kingdom)
Scotland, UK	1994-1999	6		
Spain	1995-1998	19	Gastroenteritis	Anon., 1996 Anon., 1998b
Sweden	1992-1997	350	Gastroenteritis (2)	Lindquist et al., 2000
Norway	1999	4		Unpublished data

NOTES: (1) One outbreak associated with seafood imported from Asia.
(2) One outbreak associated with consumption of crayfish imported from China.
SOURCE: European Commission, 2001.

I–2. Exposure assessment

The purpose of this exposure assessment is to quantify the exposure of consumers to pathogenic *V. parahaemolyticus* from the consumption of raw oysters in United States of America, Canada, Japan, New Zealand and Australia, using the model developed for the United States of America FDA Quantitative Risk Assessment on the Public Health Impact of Pathogenic *Vibrio parahaemolyticus* in Raw Oysters model (FDA-VPRA) (FDA, 2005). The model incorporates all phases in the harvest–post-harvest–consumption continuum, using a modular approach. Schematic representations of the pathways to be modelled are presented in Figures I–2 and I–6.

I–2.1 Microbial ecology

Vibrio parahaemolyticus is found in estuarine and coastal environments in the tropical to temperate zones. This organism is considered to be part of the autochthonous microflora in these environments and there is no correlation between the presence of this organism and faecal contamination of the environment (Kaneko and Colwell, 1977; Joseph et al., 1982). *V. parahaemolyticus* has been isolated from seawater, sediment, marine animals, plankton, and various fish and shellfish species (Joseph et al., 1982). The organism has been isolated from a number of fish species and is associated primarily with the intestinal contents (Nair et al., 1980).

Thus, *V. parahaemolyticus* is naturally present in shellfish growing and harvesting areas. Certain areas may have more favourable environmental conditions that support establishment, survival and growth of the organism, such as temperature, salinity, zooplankton, tidal flushing (including low-tide exposure of shellfish) and dissolved oxygen (Amako et al., 1987; Garay et al., 1985; Kaneko and Colwell, 1977; Venkateswaran et al., 1990). In temperate waters, *V. parahaemolyticus* is often detected in warmer months and the organism has been reported to survive in the sediment during winter (Kaneko and Colwell, 1977). However, in tropical waters, *V. parahaemolyticus* can be detected throughout the year (Natarajan et al., 1980; Deepanjali et al., 2005). While salinity and temperature are considered important factors influencing the prevalence and levels of *V. parahaemolyticus* in temperate waters (Kaneko and Colwell, 1977; DePaola et al., 2003), salinity appears to be the major factor in tropical waters (Deepanjali et al., 2005). *V. parahaemolyticus* can grow in sodium chloride concentrations ranging from 0.5% to 10%, with optimum levels between 1% and 3% (Colwell et al., 1984). Adsorption of *V. parahaemolyticus* on to plankton or chitin-containing materials occurs with higher efficiency under conditions of lower estuarine salinity (Kaneko and Colwell, 1975).

Several studies have been published on the concentration of *V. parahaemolyticus* in shellfish growing areas (Davis and Sizemore, 1982; Deepanjali et al., 2005; DePaola et al., 1990; Kaneko and Colwell, 1977; Kaysner et al., 1990a, b; Kaysner and Weagant, 1982; Kelly, 1999; Levine et al., 1993; Ristori et al., 2007). Oysters have been reported to harbour bacteriophages against *V. parahaemolyticus* and these bacteriophages may play a role in the abundance of the host (Comeau et al., 2005).

There are several pathways by which "new" *V. parahaemolyticus* strains may enter shellfish growing areas:

- by terrestrial and aquatic animals, some of which may harbour virulent strains and act as intermediate hosts (Sarkar et al., 1985);
- through "re-laying" shellfish; or
- by release of ballast water. Cargo vessels carry substantial quantities (10^6 L) of ballast water from the body of water where the voyage originates. While discharging ballast water indiscriminately is not considered an acceptable practice, it still has the potential to be discharged immediately prior to loading and, if present, *V. parahaemolyticus* is also released into the loading port. A report that strains of *Vibrio cholerae*, indistinguishable from the Latin American epidemic strain, were found in non-potable water taken from a cargo ship docked in the Gulf of Mexico indicates that ballast water may have been responsible for the spread of an epidemic strain of *V. cholerae* to the Gulf of Mexico (McCarthy and Khambaty, 1994). A similar mechanism could account for the spread of *V. parahaemolyticus*.

I–2.2 Growth and survival characteristics

V. parahaemolyticus is a mildly halophilic, mesophilic micro-organism and its general growth characteristics are shown in Table I–2 (ICMSF, 1996). Warmer temperatures and moderate salinity favour its survival and growth (Covert and Woodburne, 1972; Jackson, 1974; Nair et al., 1980; Zhu et al., 1992) and shellfish-borne illnesses caused by this organism occur in the warmer months. This has been observed in the United States of America, Asia and Europe (Daniels et al., 2000; Geneste et al., 2000).

Table I–2. Growth characteristics of *Vibrio parahaemolyticus.*

	Optimum	Range
Temperature (°C)	37	5 – 43
pH	7.8 – 8.6	4.8 – 11
NaCl (%)	1.5 – 3.0	0.5 – 10
Water activity (*Aw*)	0.981	0.940 – 0.996
Atmosphere	Aerobic	Aerobic – anaerobic

SOURCE: ICMSF, 1996.

Although outbreaks of foodborne disease associated with *V. parahaemolyticus* are less commonly reported in Europe, there have been a number of studies that indicate the importance of temperature in the survival and growth of *Vibrio*. In a two-year study undertaken in Italy on seawater and molluscs from the Adriatic Sea, it was found that *Vibrio* strains were most prevalent during the summer months (Croci et al., 2001). In another study, conducted in Norwegian waters, *V. parahaemolyticus* was detected only in July and August (Gjerde and Bøe, 1981).

In France, hydrobiological monitoring carried out near nuclear power plants built on the seashore showed that the most spectacular effect was on the density of vibrios. The levels were 100 times higher after the construction of the nuclear power plant than before, and vibrios were found at a level of 10^5/L in its surroundings. Furthermore, the annual decline in *Vibrio* densities during the colder months of the year was diminished and "overwintering" in sediment no longer occurred (Gregoire et al., 1993).

I–2.2.1 Growth rate

Growth of *V. parahaemolyticus* can be rapid, for example, doubling times of 27 minutes have been reported in crabmeat at both 20° and 30°C (Liston, 1974). Growth rates in a range of seafoods and tryptic soy broth with 2.5% salt (NaCl) have been recorded and summarized (ICMSF, 1996). These data indicate that moderate populations of 10^2–10^3 organisms/g on seafood can increase to $>10^5$ organisms/g in two to three hours at ambient temperatures between 20° and 35°C (ICMSF, 1996).

Miles et al. (1997) modelled the growth rate of *V. parahaemolyticus* based on studies of four strains at different temperatures and water activity. For each combination of temperature and water activity, bacterial growth was modelled using the Gompertz function, a sigmoid growth curve with a growth rate (slope) monotonically increasing to a maximum before falling to zero as the bacterial population reaches a steady state. The maximal rate of growth (μ_m) is the most relevant summary of the fit because the growth rate approaches its maximum rapidly and does not decline significantly until a steady state is reached.

A secondary, square root, model was used to estimate the effect of environmental parameters on the maximal growth rate:

$$\sqrt{\mu_m} = \frac{b*(T-T_{\min})*\left[\{1-\exp(c*(T-T_{\max}))\}*\sqrt{(a_w-a_{w,min})*[1-\exp(d*(a_w-a_{w,\max}))]}\right]}{\sqrt{\ln(10)}}$$

where:

μ_m = maximal growth rate ($\log_{10}$ per minute),
a_w = water activity,
T = temperature (in degree Kelvin),
b, c and d are coefficients to be fitted, and
$a_{w,min}$ and $a_{w,max}$ = theoretical lower and upper water activity limits.

Based on the data from the fastest growing strain, the estimates of the parameters were: $b = 0.0356$; $c = 0.34$; $T_{min} = 278.5$; $T_{max} = 319.6$; $a_{w,min} = 0.921$; $a_{w,max} = 0.998$; and $d = 263.64$.

The parameters T_{min}, T_{max}, $a_{w,min}$ and $a_{w,max}$ describe the range of temperature and water activity over which growth can occur. Miles et al. (1997) validated their model by comparing predicted growth with observed rates in eight other studies of growth in broth systems obtained from the literature.

A plot of the resulting model prediction for μ_m as a function of either temperature or water activity is a unimodal function with a maximum value and zero growth rate outside of the predicted range of temperatures and water activity favourable for growth. To use this equation as a prediction of growth rate in oysters, it was assumed that water activity of oysters does not vary substantially. Accordingly, this parameter was set at the optimal value of 0.985 predicted for the broth model system, where the predicted growth rate in broth at 26°C is 0.84 $\log_{10}$ per hour. This is four times greater than the rate of growth observed for *V. parahaemolyticus* in oysters held at 26°C (Gooch et al., 2002). Based on this observation, the predicted growth rate in oysters at temperatures other than 26°C was obtained by dividing the predicted rate in broth by four.

It was assumed that the growth rate in oysters is a constant fraction of the growth rate in broth at all temperatures. This assumption in the risk assessment is accounted for as an uncertainty parameter varying according to a triangle distribution in the range of 3 to 5, with a mean of 4. This evaluates the sensitivity of the conclusions to the magnitude of the relative growth rate in oysters *versus* the broth model, but does not address completely the uncertainty in so far as it is conceivable that the relative growth rate could be temperature dependent.

After harvesting, changes in the oyster are gradual and are unlikely to induce a lag phase in growth of *V. parahaemolyticus*. Consequently, for oysters, the extent of growth over time was assumed to occur at the maximum specific growth rate predicted from the average temperature, until the stationary phase is reached. This can be modelled using a model of the form:

$$\log_{10}(N(t)) = \min\{\log_{10}(N(0)) + \mu_m * t, A\}$$

analogous to the model of Einarsson (1994) or to the three-phase log-linear model (Buchanan et al., 1997), but with no lag phase, where $N(t)$ refers to the bacterial density at a given time (t) post-harvest, A is the logarithm of the maximum attainable density of *V. parahaemolyticus* in oysters, and the parameter μ_m is a function of ambient temperature and water activity, as described above.

At 26°C, the density of *V. parahaemolyticus* in oysters was observed to approach a plateau of approximately 6.0 $\log_{10}$ per gram after 24 hours (Gooch et al., 2002). We have assumed this value for the maximal density (A) at all temperatures. Figure I–1 shows predictions of the $\log_{10}$ increase in *V. parahaemolyticus* density from an initial level of 1000/g as a function of time for three ambient temperatures (20°, 26° and 32°C) at the optimal water activity.

Ideally, the average temperature used to determine the parameter μ_m in the above equation is the temperature of oyster meat of shellstock. Clearly, the temperature of oyster meat depends on the temperature of both the air and water at the time of harvest. Temperature of the oyster meat after harvest will gradually equilibrate with the temperature of the air, which has been used as a surrogate for oyster meat temperature.

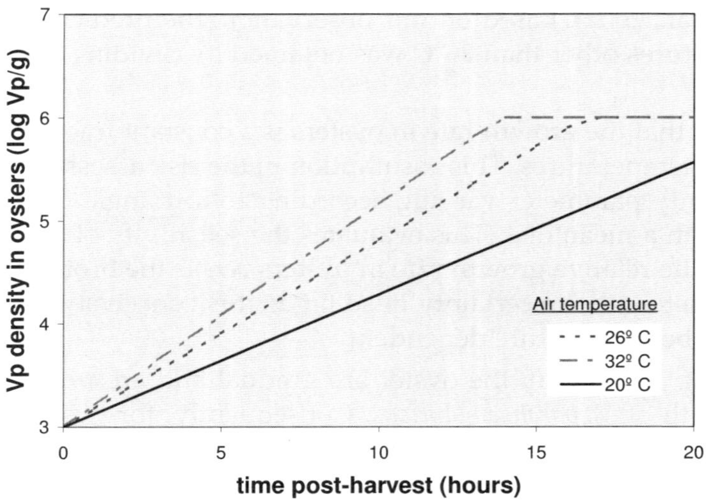

Figure I-1. Predicted loglinear growth of *V. parahaemolyticus* (Vp) from an initial density of 1000 (3 $\log_{10}$) *V. parahaemolyticus*/g as a function of ambient air temperature.

I-2.2.2 Death and inactivation

Although the ecology of *V. parahaemolyticus* has been studied (Joseph et al., 1982; Kaneko and Colwell, 1977), little is known about the growth and survival of *V. parahaemolyticus* in shellstock oysters (Cook and Ruple, 1989). By contrast, post-harvest growth of *V. vulnificus* in oyster shellstock (Cook, 1994, 1997a) and the effectiveness of various mitigation strategies for reducing *V. vulnificus* have been studied more extensively (Cook and Ruple, 1992; Eyles and Davey, 1984; Motes and DePaola, 1996; Richards, 1988; Son and Fleet, 1980). These include depuration, re-laying, refrigerated storage, high pressure and mild heat treatment. There are some studies on depuration in different shellfish species and these suggest that *V. parahaemolyticus* is not effectively removed by this process, and that there may even be multiplication of the organism in some shellfish species (Son and Fleet, 1980; Croci et al., 2002).

I-2.3 Prevalence in water, sediment and shellfish

Prevalence of *V. parahaemolyticus* is associated with the presence of particulates, zooplankton and other chitin sources (Kaneko and Colwell, 1977; NACMCF, 1992; Venkateswaran et al., 1990). Several studies have shown that *Vibrio* spp. are capable of surviving and multiplying within certain protozoa, such as *Amoeba* (Abd et al., 2005; Barker and Brown, 1994). It has also been reported that *V. parahaemolyticus* "over-winters" in the sediment and is absent from the water column and oysters during the winter months (Joseph et al., 1982; Kaysner et al., 1989; HHS, 1995). During the summer, shellfish often have levels of *V. parahaemolyticus* two hundred times greater, on average, than those in the water (DePaola et al., 1990; Kaysner et al., 1990a) suggesting that sediment should be monitored during the winter and shellfish meat during the summer. Under extreme environmental conditions, *Vibrio* species, including *V. parahaemolyticus*, may enter a "viable but non-culturable"

(VBNC) phase in marine waters and could be missed by traditional cultural methods (Bates et al., 2000; Colwell et al., 1985; Oliver, 1995; Xu et al., 1982). This issue remains controversial. Methods using gene probes are capable of detecting most virulent strains and could be useful in monitoring programmes (Gooch et al., 2002). However, some of the virulence genes, like the *trh* gene, may be present in other vibrios, such as *V. alginolyticus* and *V. harveyi* (Gonzalez-Escalona et al., 2006; Masini et al., 2007).

Micro-organisms are incorporated into shellfish by filter feeding, and factors that favour active filter feeding increase the uptake of the pathogen (Murphree and Tamplin, 1991). Shellfish species and physiology (e.g. sexual maturity, immune function, metabolic state) can affect survival and growth of disease-causing *Vibrio* spp. within shellfish (Fisher and Di Nuzzo, 1991; Kothary et al., 1997; La Peyre and Volety, 1999; Ordas et al., 1998; Volety et al., 1999). There also appears to be seasonal differences in the oyster cellular defence system, with the bactericidal activity of haemocytes (oyster blood cells) being greater in summer than in winter (Genthner et al., 1999).

The oyster parasite *Perkinsus marinus* also plays a role in the affinity of bacteria for oyster tissue and in the ability of oyster haemocytes to kill the internalized organisms (Kothary et al., 1997; La Peyre and Volety, 1999; Tall et al., 1999). Factors, such as spawning or adverse environmental conditions (including tributyltin oxide, polycyclic aromatic hydrocarbons or wood preservative leachates) reduce or stop filter feeding in shellfish. Selective feeding (e.g. new nutrient sources) may prevent or delay assimilation of *V. parahaemolyticus* into these shellfish by affecting oyster physiology and oyster-bacterial interactions (Sujatha et al., 1996; Weinstein, 1995; Wendt et al., 1996).

It is not known whether virulent and non-virulent strains are similarly affected by environmental and other factors. The presence of the urease gene may provide a competitive environmental advantage over other strains by allowing access to a wider range of nutrients (Abbott et al., 1989). Urease-positive strains have been identified as a predominant cause of *Vibrio*-associated gastroenteritis on the West Coast of the United States of America and in Mexico (Abbott et al., 1989). The presence of pathogenicity islands (physical groupings of virulence-related genes) in *V. parahaemolyticus* may foster rapid micro-evolution, promote growth and survival and result in transmission of factors (including virulence) to other strains by horizontal gene transfer (Frischer et al., 1990; Ichige et al., 1989; Iida et al., 1998). Bacteriophages may also genetically alter vibrios (Baross et al., 1978; Hurley et al., 2006; Ichige et al., 1989; Wang et al., 2006).

The distribution and variation in numbers of virulent *V. parahaemolyticus* in oysters and among oyster growing areas may need to be determined before harvest because many of the described factors may have contributed to higher concentrations of virulent strains in certain areas. During the 1998 outbreaks in the United States of America, *V. vulnificus* in shellfish harvested from the Hood Canal area of Washington State in the Pacific Northwest were responsible for 67% (32/48) of the illnesses in that state (Thieren, 1999). In the Gulf Coast area of the United

States of America, two-thirds (20/30) of the harvest sites were implicated, while in the Atlantic Northeast, only one harvest area, Oyster Bay Harbour, was implicated in the outbreak in that region (CDC, 1999).

The analysis of 141 samples of shellfish, including oysters (*Crassostrea gigas*), mussels (*Mytilus galloprovincialis*) and clams (*Ruditapes decussatus*), showed that *V. parahaemolyticus* occurred at a frequency of 11.3%, while that of *V. vulnificus* was 1.4%. Among the isolates of *V. parahaemolyticus*, 15 were recovered from oysters and one from clams (Bouchriti et al. 1995).

I–2.4 Consumption of oysters

Anyone who consumes shellfish raw is "at risk" of infection by *V. parahaemolyticus*, and the characteristics of the host are addressed in more detail in Section I–3.1. Intake data for molluscan shellfish are available from a number of governmental and non-governmental sources, but there is nevertheless a scarcity of such consumption data, as noted recently in Europe (European Commission, 2001). Also, because raw shellfish is not a commonly consumed food in many countries, for example in the United States of America, where approximately only 10 to 20% of the population will consume shellfish raw at least once during a year, some of the data are typically based on very few eaters reporting consumption. The United States Department of Agriculture (USDA) Continuing Survey of Food Intake by Individuals (CFSII) (USDA, 1989–1992) and the food frequency survey conducted by the Market Research Corporation of America (Degner, 1998) suggest that, in the United States of America, raw oysters are consumed on average approximately once every 6 weeks. The mean mass of raw oysters consumed at a single serving is 196 g, or approximately six raw large oysters (TAS, 1995). The distribution of shellfish intake will be derived from food intake surveys, food frequency surveys and from reported landings of shellfish and industry estimates of the percentage of shellfish consumed raw.

I–2.5 Modelling exposure to V. parahaemolyticus

I–2.5.1 Approaches

Four factors were used to model exposure:

- level of pathogenic *V. parahaemolyticus* in seafood at harvest;
- effect of post-harvest handling and processing;
- ability of the organism to multiply to an infective dose; and
- number of pathogenic *V. parahaemolyticus* consumed.

In Asia, *V. parahaemolyticus* is a common cause of foodborne disease. In general, the outbreaks are small in scale, involving fewer than 10 cases, but occur frequently. Prior to 1994, the incidence of *V. parahaemolyticus* infections in Japan had been declining; however, in 1994-95 there were 1280 reports of infection due to the organism (Anon., 1999a) and, during this period, *V. parahaemolyticus* food poisonings outnumbered those of *Salmonella* food poisoning. From 1996 to 1998, there were 496 outbreaks and 24 373 cases of *V. parahaemolyticus* reported, while from 1999 to 2005,

25 211 cases were reported (data from Ministry of Health, Labour and Welfare, Japan). In general, outbreaks were more prevalent in the summer, with a peak in August. Boiled crabs caused one large-scale outbreak, involving 691 cases. The increased incidence during 1997–1998 has been attributed to an increased incidence of serovar O3:K6.

Between 1986 and 1995, some 197 outbreaks of foodborne disease were caused by *V. parahaemolyticus* in Taiwan (Pan et al., 1997), while in 1997 over 200 outbreaks were reported, including an outbreak of 146 cases acquired from boxed lunches (Anon., 1999b).

During 1997 and 1998 there were more than 700 cases of illness due to *V. parahaemolyticus* in the United States of America, the majority of which were associated with the consumption of raw oysters. In two of the 1998 outbreaks, a serotype of *V. parahaemolyticus,* O3:K6, reported previously only in Asia, emerged as a principal cause of illness for the first time. Subsequent studies on these strains have revealed pandemic spread. It was suggested that warmer than usual water temperatures were responsible for the outbreaks

In Europe, few data exist on the incidence of *V. parahaemolyticus* infections, one reason being that such infections are not notifiable. However, the current knowledge of the incidence in Europe has been presented in *Opinion of the Scientific Committee on Veterinary Measures relating to Public Health on* Vibrio vulnificus *and* Vibrio parahaemolyticus *in raw and undercooked seafood* (European Commission, 2001).

As a consequence of the data patterns, the exposure assessment was divided into two separate modules: a Harvest Module; and a Post-harvest Module that includes retail and consumption data.

Factors influencing the risk of illness posed by *V. parahaemolyticus* were identified and incorporated into each module as appropriate.

Water temperatures for Japanese, Australian and New Zealand oyster harvesting areas were limited. Based on available data, monthly average temperatures and standard deviations for temperatures were obtained. From these summary data, a simulation for each quarter was run using 30 random selections from each of the three-month average statistics for a total of 90 possible temperature values. The average and standard deviation of the simulation were calculated and saved. This process was repeated until 1000 "quarterly" means and standard deviations were available for use in the variability and uncertainty simulations for the growing areas selected: Hiroshima Bay in Japan, Wallis Lake in Australia and Orongo Bay in New Zealand.

I–2.5.2 Assumptions

In the present FAO/WHO risk assessment, the work undertaken to date, including the assumptions made, were reviewed by a group of experts at a joint FAO/WHO expert consultation on risk assessment of microbiological hazards in foods that was convened in July 2001 (FAO/WHO, 2001).

Harvest Module assumptions

Based on the information currently available, the presence of the thermostable direct haemolysin (*tdh*) gene was used as the basis for pathogenicity. It is not currently known what average numbers of *tdh*+ strains exist in shellfish, nationally or regionally. The estimates made in the *V. parahaemolyticus* risk assessment, based on the observed frequency of *tdh*+ isolates, were the best possible from the data currently available. However, since it is currently not known how this frequency may vary from one year to the next, a two-fold up or down triangle distribution was assumed.

Also, within a given year, there is uncertainty about the variance between the percentage of pathogenic *V. parahaemolyticus* in one composite of oysters and the next. For example, for the United States of America, with the exception of the Pacific Coast (where the range was 2% to 4%), the percentage of pathogenic *V. parahaemolyticus* in a given year ranged from 0.1% to 0.3%. However, these estimates are based on older data and may not be predictive of future years, given that the frequency of percentage of pathogenic *V. parahaemolyticus* may be changing as new outbreak strains emerge or re-emerge, such as the emergence of O3:K6, or recurrence of known outbreak strains, such as O4:K12. It has also been noted that the proportion of pathogenic strains occurring can vary from region to region. For example, the above strains have tended to occur with greater frequency in Asia than in the United States of America (FAO/WHO, 2001).

Post-harvest Module assumptions

Several assumptions were made based on the knowledge of current post-harvest practices and information available in each country. The time oysters are harvested to the time they are refrigerated was based on the current National Shellfish Sanitation Program (NSSP) requirements (ISSC and FDA, 1997), put into effect in 1997 in the United States of America. The extent of growth that occurs during the period from harvest until oysters are first placed under refrigeration is determined by three factors:

- growth rate of *V. parahaemolyticus* as a function of temperature;
- temperature of oyster meat after harvest; and
- length of time held unrefrigerated.

The growth rate of pathogenic *V. parahaemolyticus* in oysters was assumed to be one fourth that in broth culture at all temperatures. Also, since the *V. parahaemolyticus* organisms do not change their growth environment after harvest (within the oyster meat), it was assumed that lag time was negligible and it was therefore omitted from the growth model. Regarding growth rates, preliminary studies at the Gulf Coast Seafood Laboratory (GCSL) in the United States of America showed no significant difference between pathogenic and non-pathogenic strains of *V. parahaemolyticus.*

Since data on cooling rates of commercial oyster shellstock have not been located to date, the time for oysters to cool after being placed under refrigeration was assumed to vary. Cooling rates are dependent on the efficiency of the cooler, the quantity of oysters to be cooled and their arrangement in the cooler. A uniform distribution between 1 and 10 hours was used to model this parameter. This was based on preliminary experiments carried out at GCSL in the United States of America for the time it took a single shell oyster at 30°C placed into a 3°C cooler to reach that temperature, and the time it took for 24 oysters in an uninsulated plastic container at 26°C to reach 3°C. Cooling times in other countries are based on anecdotal information from industry, and it must be noted that in some countries refrigeration *per se* is not used. For example, in the Australian location used in this assessment, oysters are stored in a building that may have air conditioning rather than refrigeration.

For the sake of simplicity of the model, it was assumed that consumption patterns were the same for both the sensitive and otherwise healthy population, for all regions in each country. It was assumed that all virulent or pathogenic strains of *V. parahaemolyticus* are equally virulent with the same dose-response as those strains fed to human volunteers in earlier studies. This assumption was based on personal communication with Mitsuaki Nishibuchi, Kyoto University, Japan, who stated that, due to lack of information, it is not known whether there are differences in virulence among different strains.

I–2.6 Harvest module

Although a number of factors have been identified as potentially affecting the numbers of pathogenic *V. parahaemolyticus* in oysters at time of harvest, there are not sufficient quantitative data available to incorporate all of these factors into a predictive model. Incorporation of an environmental factor into the simulation as a predictor of *V. parahaemolyticus* numbers at harvest requires both the relationship between *V. parahaemolyticus* numbers to the parameter of interest, and the regional and temporal variation of that parameter within the environment to be identified. Moreover, due to the relatively low prevalence of pathogenic *V. parahaemolyticus* and limitations of current methods of detection, the distribution of pathogenic *V. parahaemolyticus* is not well understood. A critical issue in the development of the Harvest Module simulation is the use of the estimated distribution of total *V. parahaemolyticus* numbers to bridge this data gap and to derive an estimate of the distribution of pathogenic *V. parahaemolyticus* numbers in oysters at harvest. Figure I–2 illustrates the parameters considered in modelling the Harvest Module.

The main factors that have been identified as potentially affecting the numbers of pathogenic *V. parahaemolyticus* in oysters at the time of harvest are water temperature and salinity, and these are addressed in more detail below. Preliminary analyses indicated that salinity is not as strong a determinant of *V. parahaemolyticus* numbers as water temperature, and therefore is represented as a dotted bubble in Figure I–2.

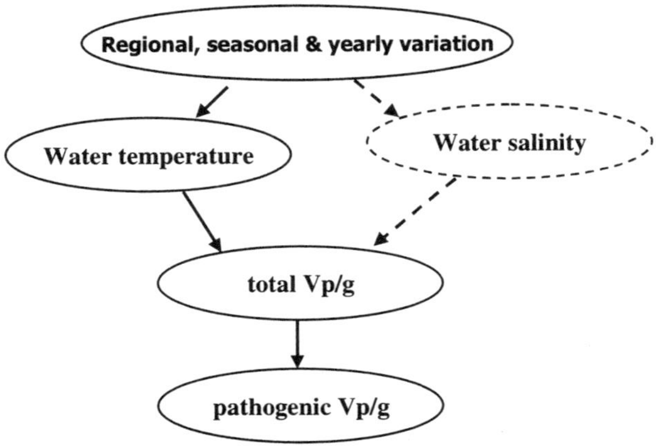

Figure I–2. Schematic depiction of the Harvest Module of the *V. parahaemolyticus* (Vp) risk assessment model.

I–2.6.1 Effect of water temperature and salinity on prevalence

A number of studies have been carried out on the prevalence of *V. parahaemolyticus*, but many of these report only presence or absence of *V. parahaemolyticus*, making them of limited value for quantitative risk assessment (Hariharan et al., 1995; Kelly and Stroh, 1988a; Kiiyukia et al., 1989). In the United States of America, some studies that did measure *V. parahaemolyticus* numbers in oysters were also limited because samples were either obtained from a single estuary (Chan et al., 1989; Hariharan et al., 1995; Kaysner et al., 1990b; Kelly and Stroh, 1988a, b; Kiiyukia et al., 1989; Tepedino, 1982), were not seasonal (Chan et al., 1989; Kaysner et al., 1990b; Tepedino, 1982) or did not report salinity and temperature (Chan et al., 1989; Tepedino, 1982). A number of there are summarized in Table I–3.

There are limitations to the methodology currently used in the enumeration of *V. parahaemolyticus*. In a study that evaluated four methods for enumerating the organism in natural seawater and oysters, it was found that there was considerable variability between methods for *V. parahaemolyticus* recoveries, with highest recoveries being obtained using filtration through a hydrophobic grid membrane (DePaola et al., 1988). In a subsequent study, DePaola et al. (1990) used the hydrophobic grid membrane filtration (HGMF) procedure developed by Watkins et al. (1976) and later revised by Entis and Boleszczuk (1983).

Table I–3. Surveys on the numbers and prevalence of *V. parahaemolyticus* (Vp) in oysters, sediment and water

Location	Period	Samples taken	No. of Vp	Prevalence	Environment	Ref
British Columbia, Canada	July – August March – April	Cultivated oyster Natural oyster Cultivated oyster Natural oyster	—	21% 44% nd nd	Estuarine waters	Kelly and Stroh, 1988a
British Columbia, Canada	Summer	Estuarine water	70 cfu/mL	11-33% of water samples	Estuarine water – warm	Kelly and Stroh, 1988b[(1)]
Willapa Bay, Washington State, USA	August	Water Sediment Oyster	0.5–3.0 MPN/g 1.6–5.4 MPN/g 1.5–4.0 MPN/g	—	Salinity 23.6–30.5 ppt Temperature 15.5–22.6°C	Kaysner et al., 1990b
Long Island, USA	October to June	Oysters	3.6–23 MPN/g.	33%	-	Tepedino, 1982
Pr. Edward Island, Canada	All year	Oysters Mussels	—	6.7% 4.7%	—	Hariharan et al., 1995
Hong Kong	June through October	oysters (harvest) mussels (market) clams (market)	3.4×10^4 4.6×10^4 6.5×10^3/g	—	—	Chan et al., 1989
Japan	Spring, summer and autumn	Water Sediments Market oyster	0.66–4.93 MPN/100 mL 0.2–2.4 MPN/100 g —	2/8	Temperature 9.9–29.5°C Salinity 24% (4 sampling areas). 1 estuarine area <16 – >24%	Kiiyukia et al., 1989
Japan		Clams	<3–>10^5 MPN/10 g	—	—	Hara-Kudo et al., 2003
Hiroshima Bay, Japan	July 1987 to June 1988	Oysters	10^3–10^1/100 g	69% (May to October)	Temperature ranged from 19.3–22.0°C	Ogawa et al., 1989
Mulky estuary, India	All year	Oysters (*Crassostrea madrasensis*)	Total Vp: 10^2–10^4/g Pathogenic Vp: 10^1–10^2/g	93.8% 10.2%	Temperature from 25–32°C Salinity from 8.3 to 30.5 ppt	Deepanjali et al., 2005

NOTES: nd = none detected. (1) Reported an association with *V. parahaemolyticus* illness and *V. parahaemolyticus* density in the estuarine waters of British Columbia.

When all suspect colonies were tested for confirmation, the precision of the HGMF method was shown to be greater than the three-tube most probable number (MPN) procedure (Entis and Boleszczuk, 1983; Watkins et al., 1976).

As the basis for the present study, the relationship between total *V. parahaemolyticus* densities in oysters and water temperature was quantified using three comprehensive survey data sets: DePaola et al. (1990); FDA/ISSC (2001); and Washington State Department of Health (2000, 2001). These data sets were selected for quantitative modelling because they measured levels of *V. parahaemolyticus* in oyster meat and water temperature over all seasons.

Because different methodologies were used for enumeration in these three surveys, the data sets were not pooled. Instead, regression models were fitted separately to each data set. A relatively large proportion of samples within the data sets had non-detectable levels of *V. parahaemolyticus*. In DePaola et al. (1990), 26 of 61 oyster samples (43%) did not have detectable *V. parahaemolyticus* (the lower limit of detection is approximately 10 cfu/g). In the 2001 FDA/ISSC study (later published as Cook, Bowers and DePaola, 2002), 232 of 624 (37%) samples analysed for total *V. parahaemolyticus* were found to have less than the limit of detection (10 cfu/g) and 93 of 262 (36%) of oyster samples were below the limit of detection (0.3 cfu/g) in the Washington State monitoring data (Washington State Department of Health, 2000; 2001). For regression analysis, it was assumed that *V. parahaemolyticus* was present in these non-detect samples at levels below the detection limit (i.e. the true density was below the limit of detection), but never zero.

Regression analysis was carried out on *V. parahaemolyticus* densities in oysters from the three studies. The Tobit regression is a maximum likelihood procedure with likelihood reflecting both the probability of obtaining a non-detectable outcome at a given temperature as well as the probability distribution of observable numbers given that a sample has detectable cells of *V. parahaemolyticus*. The effect of this likelihood structure is to weight the influence of non-detection on estimated trends differently from samples with quantifiable numbers. The influence of non-detection is based on the probability of the number of cells in a sample falling below a fixed limit of detection rather than the assumption that a non-detectable measurement corresponds to an observed and quantifiable number at the limit of detection, or one-half the limit of detection, as is commonly assumed.

Plots of the best fitting regression line versus temperature and the associated 5th and 95th percentile confidence intervals are shown in Figures I–3 to I–5 for each of the three data sets. In these figures, non-detectable *V. parahaemolyticus* levels were replaced with randomly imputed values (open circles) based on the maximum likelihood estimate (MLE) of the regression relationship. Regression analysis of the three data sets indicated that the effect of temperature on the mean $\log_{10}$ total *V. parahaemolyticus* densities was approximately linear in the range of water temperatures sampled.

Results of the Tobit regression analysis of the three data sets were used to generate 1000 sets of parameters for the relationship of water temperature to total *V. parahaemolyticus* densities in oysters. These sets of regression parameters were used to represent uncertainty of the water temperature relationship and variance of total *V. parahaemolyticus* densities in the Monte Carlo simulations. For the Gulf Coast, Mid-Atlantic and Northeast Atlantic regions, the uncertainty from the regression analyses shown in Figures I–3 and I–4 were used. Approximately 500 sets of parameters from distributions of the model fits to these data sets were obtained and combined. The resulting 1000 sets of parameters were used once for each of the 1000 model simulations for these three regions. For the Pacific Northwest region, the 1000 parameters were obtained from the distribution shown in Figure I–5.

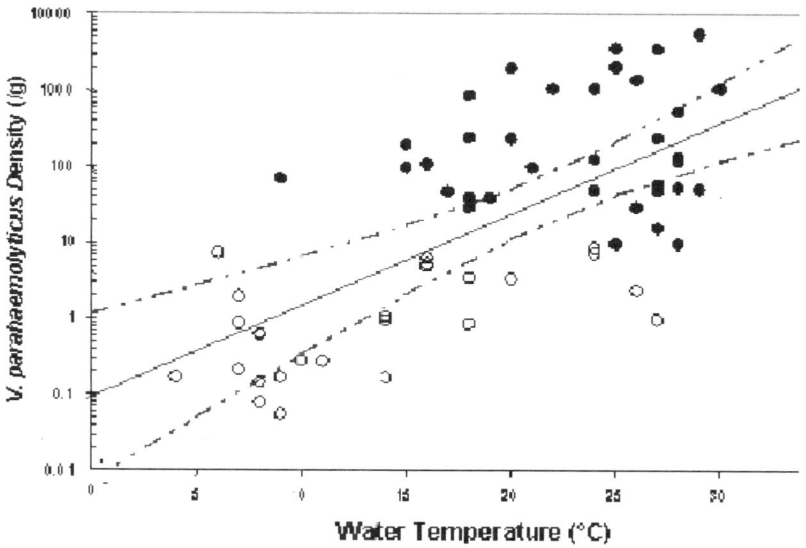

Figure I–3. Tobit regression fit of the Vibrio *parahaemolyticus* densities in oysters versus water temperature using the DePaola et al. (1990) data set.
[Solid line is the best estimate of the median *V. parahaemolyticus*/g. Dashed lines show the 5th and 95th percentile confidence limits. Closed circles are *V. parahaemolyticus* detectable values from DePaola et al., (1990). Open circles are randomly imputed values for samples with densities less than the limit of detection (10 cfu/g).]

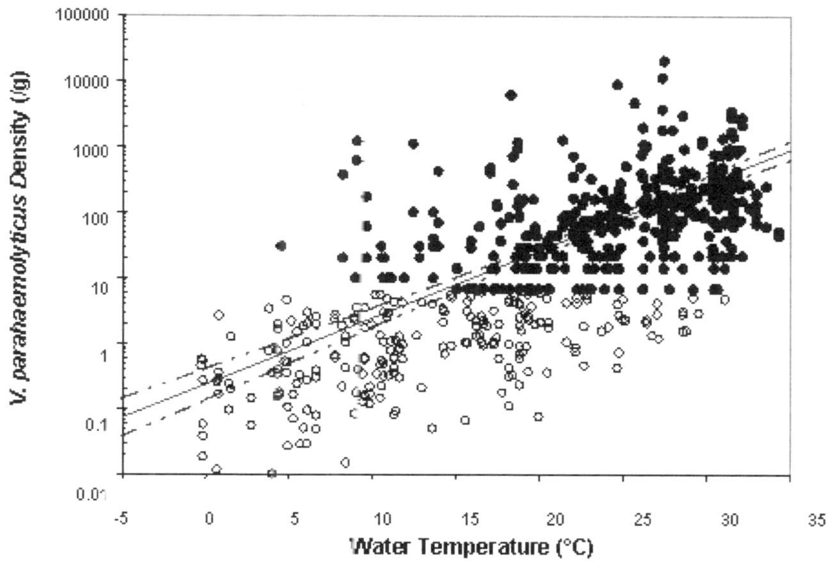

Figure I–4. Tobit regression fit of the *Vibrio parahaemolyticus* densities in oysters versus water temperature using the FDA/ISSC (2001) data set.
[Solid line is the best estimate of the median *V. parahaemolyticus*/g. Dashed lines show the 5th and 95th percentile confidence limits. Closed circles are *V. parahaemolyticus* detectable values from FDA/ISSC (2001). Open circles are randomly imputed values for samples with densities less than the limit of detection (10 cfu/g).]

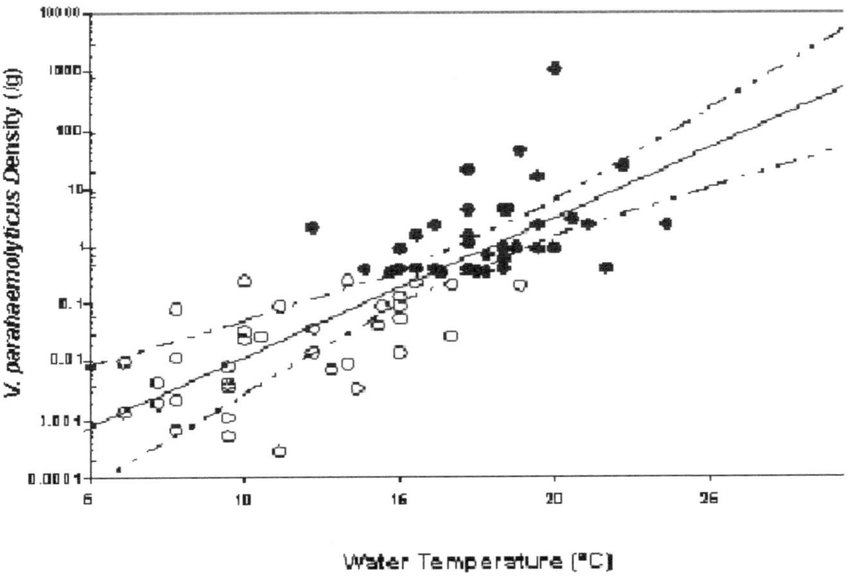

Figure I–5. Tobit regression fit of the *Vibrio parahaemolyticus* densities in oysters versus water temperature using the Washington State Department of Health (2000; 2001) data sets [Solid line is the best estimate of the median *V. parahaemolyticus*/g. Dashed lines show the 5th and 95th percentile confidence limits. Closed circles are *V. parahaemolyticus* detectable values from Washington State Department of Health (2000; 2001). Open circles are randomly imputed values for samples with densities less than the limit of detection (0.3 cfu/g).]

The effect of regression parameter uncertainty was implemented in the risk assessment by using a multivariate normal approximation for parameter uncertainty for each of the three data sets. Accounting for the effect of the uncertainty in the data sets was implemented in Monte Carlo simulations by generating a sample of 1000 sets of parameters from the uncertainty distributions. Independent estimates of method error for each of the three data sets were then used to correct this additional variance in the observed data.

Growth of *Vibrio parahaemolyticus* during intertidal exposure

In the present study, oysters in the Australian, New Zealand and Canadian growing areas were harvested by picking during the intertidal period. To model the effect of intertidal harvesting on *V. parahaemolyticus* densities, the effect of elevated oyster temperatures and duration of exposure during the collection process was modelled as a separate growth step occurring prior to growth associated with transport to processing facilities at ambient air temperature. The loglinear growth rate model described in the Post-harvest Module was used. To predict the growth of *V. parahaemolyticus* in intertidal harvested oysters prior to refrigeration, the growth rate model was applied twice. It was first applied to determine the extent of growth that corresponds to 4 to 8 hours of intertidal exposure and then to determine the extent of growth that occurs during subsequent transportation (1 hour).

In FDA (2005), the Pacific Northwest was also modelled as intertidal and the relationship of water temperature to total *V. parahaemolyticus* densities in oysters as described in Figure I–5 was used for modelling growth during intertidal harvest in

Australia, New Zealand and Canada.

In Japan, Hara-Kudo et al. (2003) determined total *V. parahaemolyticus* and proportion of *tdh+* in clams (Table I–3).

In New Zealand, prevalence and numbers of *V. parahaemolyticus* in a range of oysters from three sites in the north island are presented in Tables I–4 to I–6; in general, the organism rarely exceeded 10/g oyster meat (Fletcher, 1985). Prior to this there is a record of *V. parahaemolyticus* being isolated from shellfish in the Bay of Islands; twelve strains were isolated and serotyped as O3:K7 (four cultures), O4:K8, O4:K12 (three cultures), O4:K34, O7:K19, O8:K39 and O9:K44 (Cawley and Norris, 1973). As part of a survey under the Ministry of Health's Domestic Food Monitoring programme, *V. parahaemolyticus* was isolated from 2/8 (25%) of oyster samples from the Waikato region of New Zealand (Lake et al., 2003).

Table I–4. Surveys on the numbers and prevalence of *V. parahaemolyticus* (Vp) (mean MPN/g of meat) in oysters in New Zealand (after Fletcher, 1985).

	1981–82		1982–83		1983–84	
	Vp	**Water temp (°C)**	**Vp**	**Water temp (°C)**	**Vp**	**Water temp (°C)**
Nov	<1	19	1	18	<1	14
Dec	<1	19	<1	21	2	21
Jan	<1	23	<1	23	30	21
Jan	1	—	—	19	<1	22
Jan	10 000	—	—	—	40	22
Feb	5	—	<1	22	4	23
Mar	4	19	2	23	<1	21
Mar	2	—	4	20	<1	21
Mar	3	—	1	—	<1	—
Apr	2	17	<1	17	<1	18
May	<1	15	<1	15	<1	16
May	<1	—	<1	15	<1	15

Table I–5. *V. parahaemolyticus* (Vp) (mean MPN/g of meat) in Whangaroa and Coromandel (New Zealand) oysters (1982–83).

	Whangaroa		Coromandel	
	Vp	**Water temp. (°C)**	**Vp**	**Water temp. (°C)**
Nov	<1	13	3	18
Dec	3	20	1	18
Jan	1	—	<1	18
Jan	3	—	3	—
Jan	—	—	—	—
Feb	2	20	3	15
Mar	<1	21	3	21
Mar	<1	21	4	20
Mar	3	21	3	23
Apr	8	19	<1	14
May	8	15	3	15

Table I–6. *V. parahaemolyticus* (Vp) (mean MPN/g of meat) in Whitianga (New Zealand) oysters (1982–84).

	1982-83		1983-84	
	Vp	**Water temp. (°C)**	**Vp**	**Water temp. (°C)**
Nov	1	17	<1	15
Dec	1	17	<1	19
Jan	1	17	<1	22
Jan	<1	17	7000	—
Jan	2	—	<1	—
Feb	3	21	3	17
Mar	11	20	10	19
Mar	<1	18	2	19
Mar	3	15	—	—
Apr	8	18	—	—
May	8000	14	<1	14

In two surveys in Australia, Desmarchelier (1978) surveyed *V. parahaemolyticus* in Sydney Rock oysters (*Saccostrea glomerata* syn. *commercialis*) from 8 sites. In the first survey, 41/60 samples were positive and in the second, 128/633 samples were positive for *V. parahaemolyticus*. The author noted a direct relationship between total *V. parahaemolyticus* and temperature (Table I–7).

Table I–7. Numbers of *V. parahaemolyticus* (Vp) in Sydney rock oysters (*Crassostrea glomerata* (syn. *commercialis*) in Australia

Temp (°C)	Mean log Vp/100 g
<16	Not detected
16–20	< 1
21–24	1–2
25	3

SOURCE: Desmarchelier, 1978.

Davey et al. (1982) detected *V. parahaemolyticus* in three subsamples of undepurated oysters at a level of 4–6/g and in one subsample of depurated oysters at 0.8/g, and Eyles et al. (1985b) found 19/21 oyster meat samples were positive, with a geometric mean MPN of 7.3/g and a range of 0.3–50/g. New South Wales (NSW) Department of Health in Australia examined samples of Sydney rock oysters between April 1989 and April 1990 for *V. parahaemolyticus* (Table I–8).

Table I–8. Prevalence of vibrios in Sydney Rock oysters (*Saccostrea glomerata* syn. *commercialis*) in New South Wales, Australia.

	V. parahaemolyticus **Number positive/total**		***V. parahaemolyticus*** **Number positive/total**
1989		1990	
Jan	–	Jan	10/11
Feb	–	Feb	42/62
Mar	–	Mar	12/19
Apr	8/13	Apr	31/44
May	1/2	May	12/19
Jun	3/4		
Jul	0/1		
Aug	4/5		
Sep	–		
Oct	3/3		
Nov	5/9		
Dec	11/12		

SOURCE: John Sumner, pers. comm.

Effect of salinity

Salinity can influence the prevalence and growth of *V. parahaemolyticus* in oysters, and preliminary modelling included a consideration of that parameter (see FDA, 2005). Subsequent consideration of the model indicated that water salinity is not as strong a determinant of *V. parahaemolyticus* levels in the regions that account for essentially all the commercial harvest, and was overshadowed by the impact of water temperature (see Appendix 5 of FDA-VPRA, FDA, 2005). For these reasons, salinity was not a "modelled effect" *per se* in the regression. However, since Australia, New Zealand and Canada were all modelled as intertidal based on the Pacific Northwest, which has relatively high salinity, its effect at intertidal harvest was included *de facto* in the modelling.

Table I–9 presents levels of *V. parahaemolyticus* from six growing areas in Canada, which illustrate that high levels ($>10^3$/g) can occur in summer months (Klaus Schalle, Canadian Food Inspection Agency, pers. comm.).

Table I–9. *V. parahaemolyticus* (mean MPN/g) levels in Canadian oysters

Harvest Location	Statistical Area	Harvest Date	Harvest Time	Product Type/Class	Mean MPN/g
Gorge Harbour	13-15	2002/05/14	9:30	Oysters – Longline	3
	13-15	2002/06/03	8:45	Oysters - Tray	3
	13-15	2002/06/16	9:30	Oysters - Tray	3
	13-15	2002/07/02	9:15	Oysters - Tray	3
Ship's Point	14-8	2002/05/21	8:10	Oysters - Beach	3
	14-8	2002/06/03	7:45	Oysters - Beach	3.6
	14-8	2002/06/17	6:25	Oysters - Beach	110
	14-8	2002/06/25	11:25	Oysters - Beach	9.2
	14-8	2002/07/02	12:10	Oysters - Beach	
Twin Island	15-3	2002/05/21	7:00	Oysters - Beach	3
	15-3	2002/06/04	7:00	Oysters - Beach	3
	15-3	2002/06/18	7:00	Oysters - Beach	430
	15-3	2002/07/02	7:00	Oysters - Beach	
Sykes Island	16-12	2002/05/13	11:20	Oysters - Tray	3
	16-12	2002/05/27	9:45	Oysters - Tray	3
	16-12	2002/06/10	9:00	Oysters - Tray	3
	16-12	2002/06/24	10:15	Oysters - Tray	430
Ladysmith Harbour	17-7	2002/05/13	10:50	Oysters - Beach	3
	17-7	2002/05/27	9:20	Oysters - Beach	3
	17-7	2002/06/10	9:45	Oysters - Beach	15
	17-7	2002/06/24	9:00	Oysters - Beach	4600
	17-7	2002/07/02	15:40	Oysters - Beach	2400
Ritherdon Bay	23-3	2002/05/13	11:37	Oysters - Beach	3
	23-3	2002/05/28	16:00	Oysters - Beach	3
	23-3	2002/06/10	12:10	Oysters - Longline	3
	23-3	2002/06/25	18:00	Oysters - Beach	3.6

In growing areas of Australia and New Zealand, rainfall data or salinity data are used as a regulatory mechanism at harvest to manage levels of bacteria and viruses of faecal origin. Most growing areas are classified as "Conditionally Approved", which means that they are closed when there are conditions that may bring pollution to the area. The most common management system is rainfall: after a specified amount of rainfall at the official rain gauge the area is closed for a specified number of days. In others, where salinity has been correlated with rainfall, readings on salinity gauges are used for managing harvesting. For example, in New Zealand, the Orongo Bay area is one of two growing areas managed by salinity criteria (D.J. McCoubrey, pers. comm.). Harvesting ceases when a specified salinity level (<17 ppt) is reached. The rate of change of salinity during tidal cycles may also be used for determining closing and opening of harvest areas. The 24-hour mean salinity must exceed 23 ppt before an area can be considered for re-opening, but the time lapse between reaching this level and and allowing re-opening will vary depending on the minimum salinity levels that occurred.

In Australia, harvesting is halted if salinity falls to <17 ppt. In practice, because of the time lapse before advising growers that the pollution event is over, harvest does not re-commence until salinity is 22–23 ppt.

I–2.6.2 Water Temperature Distributions

In the FDA-VPRA (FDA, 2005), regional and seasonal distributions of water temperatures were developed based on accumulated records from coastal water buoys (National Buoy Data Center data – NBDC). Seasons were defined by calendar month: winter (January–March), spring (April–June), summer (July–September) and autumn (October–December). For each region and season a shallow water buoy was selected as representing the water temperature distribution for oyster harvest areas within that region+season combination.

Because oyster harvesting in the United States of America outside of the Pacific Coast region commences early in the morning and ends mid- or late afternoon, the daily water temperature recorded at noon was considered to represent an average daily temperature. The distribution of these "average" temperatures within a given region and season varies from year to year, with wider variations occurring during the transitional seasons of spring and autumn.

In the Australian and New Zealand sites selected for this assessment, oysters are grown in the intertidal area. Oyster spat is collected naturally on sticks placed in the marine environment and the sticks placed on racks to on-grow the oysters. The sticks are placed about 250 mm apart on racks. Sometimes oysters are split off the sticks and placed in netlon bags or plastic bins to be on-grown to larger sizes. Growing in bags also allows for better control of the shell shape. Whether grown on sticks or in bags, oysters take about 12 to 18 months to mature to a harvestable size.

The main oyster harvest season is May–November, as during the colder months they are in peak condition. Spawning is in late December to January, depending on water temperatures. However, oyster harvesting does occur throughout the year in many areas as there are markets both in New Zealand and overseas that accept oysters in the poorer, after-spawning condition. In New Zealand, mean salinity, water and air temperatures typical of growing areas in the north of the country are presented in Table I–10 (D.J. McCoubrey, New Zealand Food Safety Authority, pers. comm.).

Table I–10. Mean salinity, water and air temperatures during 2001, Orongo Bay, New Zealand.

	Water (°C)	Air (°C)	Salinity (ppt)
Jan	21	29	35
Feb	21	28	35
Mar	21	26	35
Apr	20	24	35
May	17	21	35
Jun	14.5	20	34
Jul	14.5	19	34
Aug	14.5	18	34
Sep	15	20	34
Oct	17.5	23	33
Nov	18	23	34
Dec	21	27	35

In Australia, the Sydney Rock oyster is grown in the intertidal zone. Oyster spat are placed on racks to on-grow the oysters. The main oyster harvest season coincides with Christmas and Easter. For Australia, salinity and water temperature, based on NSW government data, are presented for the Wallis Lake growing area (Table I–11).

In United States of America, within a given year, the distribution of the noon water temperature was found to be unimodal within a given range. This empirical distribution is adequately approximated as a normal distribution provided that no weight is given to implausible values outside the historical range of values that may be expected. Differences in these distributions from one year to the next are evident in the buoy data. This year-to-year variation in the water temperature distributions has been characterized by calculating the central tendency and variation in both the mean and standard deviation of these distributions (Table I–12).

NBDC measures surface water temperature (sensors are generally 1.0 to 1.5 m deep). There were no near-shore NBDC buoys recording water temperatures in oyster growing areas and consequently, for this region, seasonal and year-to-year variations in water temperature distributions were developed based on compiled data from Washington State shellfish specialists using Washington State Department of Health information from 1988 to 1999.

Table I–11. Salinity and water temperatures at Wallis Lake, NSW, Australia.

Year	Month	Temperature (°C)			Salinity (ppt)			Samples (no.)
		Mean	Max	Min	Mean	Max	Min	
2001	April	23.6	26.5	22.1	32.4	34.2	27.8	20
	May	19.6	20.7	18.6	21.8	31.7	9.3	8
	June	19.1	20.8	16.8	31.3	34.8	26	16
	July	17.7	19.2	15.1	32.8	35.1	30.2	24
	August	17.3	19.8	15.7	34.1	35.2	32	20
	September	19.4	21.9	17.5	33.8	35	32.2	18
	October	20.1	21.6	17.4	33.4	35.5	32.3	35
	November	22.5	25.8	20.3	33.8	36	18.4	56
	December	24.6	28.9	22.1	34.5	36.7	31.8	35
2002	January	24.8	26.9	23.3	34.5	36.5	32.8	35
	February	23.9	25.1	20.9	28.7	34.6	9.9	49
	March	25.2	26.7	24.2	33.5	34.7	30.9	28
	April	21.3	23.9	18.3	18.3	29.3	29.3	35

Table I–12. Summary statistics of the year-to-year variation in the mean and standard deviation of noon water temperature distributions for different regions and seasons in the United States of America.

Region	**Mean (standard deviation) water temperature** (°C)			
	Winter (Jan–Mar)	**Spring** (Apr–Jun)	**Summer** (Jul–Sep)	**Autumn** (Oct–Dec)
Northeast Atlantic	4.51 (1.23)	12.0 (4.2)	20.7 (1.34)	12.0 (3.37)
Mid-Atlantic	3.92 (1.92)	16.8 (5.1)	25.0 (1.8)	11.6 (5.1)
Gulf Coast	14.2 (2.7)	24.5 (3.5)	28.9 (1.5)	17.9 (4.5)
Pacific Northwest	8.1 (1.62)	13.7 (2.4)	17.4 (2.4)	10.7 (2.8)

SOURCE OF DATA: National Buoy Data Center (NBDC) (http://www.ndbc.noaa.gov/) and supplied by N. Therien, Washington State shellfish specialist, pers. comm., 1999.

I–2.6.3 Prediction of the distribution of pathogenic *V. parahaemolyticus* numbers

Estimates of the percentage of total *V. parahaemolyticus* isolates that have been found to be pathogenic in several studies in the United States of America are presented in Table I–13, from which it can be seen that differences were observed in the various United States of America regions, with higher percent pathogenic values observed in the Pacific Northwest compared to the Gulf Coast and Atlantic regions. The studies of DePaola et al. (2002) and Kaufman et al. (2003) were considered the most appropriate for estimating the ratio of pathogenic to total *V. parahaemolyticus* in oysters. The data from these two studies indicated that the number of pathogenic *V. parahaemolyticus* in sample portions was frequently non-detectable. In addition, high numbers of pathogenic micro-organisms were sometimes observed in samples that had low counts of total *V. parahaemolyticus* in replicate samples. Some degree of variation is expected due to the natural processes of growth and competition between different strains of *V. parahaemolyticus* in the presence of other microflora in the oysters.

While the studies by DePaola et al. (2003) and Deepanjali et al. (2005) suggest that there may be some seasonal variation in the percentage of *V. parahaemolyticus* that are pathogenic, these findings have not been replicated in other studies. Accordingly, for the purpose of this risk assessment, the ratio between pathogenic and total *V. parahaemolyticus* densities was assumed to be temperature independent.

Given the low densities of pathogenic *V. parahaemolyticus* in oysters and the resulting high frequency of non-detectable amounts in samples, distributions of percentage pathogenic were estimated based on the assumption that pathogenic counts in sample portions were distributed according to a Beta-Binomial distribution. The Beta-Binomial distribution is a flexible, two-parameter distribution commonly used to model variability of proportions. It is described fully in Appendix 5 of the FDA-VPRA (FDA, 2005).

For the other countries involved in this risk assessment, there was little or no information on prevalence of pathogenic strains of *V. parahaemolyticus*.

Table I–13. Estimates of pathogenic and total *V. parahaemolyticus* in the United States of America.

Oyster samples		*V. parahaemolyticus* isolates			
No. tested	No. pathogenic(1)	No. tested(2)	No. pathogenic(1)	% pathogenic	Region (Source of data)
153(3)	ND(4)	2 218 (MPN)	4 KP+	0.18	Gulf Coast (Thompson and Vanderzant, 1976)
60	13	5 159 (DP)	44 TDH+	0.18(7)	Gulf Coast (Kaufman et al., 2003)
198	8	3 429 (DP)	9 TDH+	0.3	Gulf Coast, Mid-Atlantic, Northeast Atlantic (FDA/ISSC, 2000; Cook et al., 2002)
106	3	5 600 (MPN+DP)	16 TDH+	0.3	Texas (DePaola et al., 2000)
156	34	6 018 (EB) 6 992 (DP)	46 31	0.76 0.44	Gulf Coast (DePaola et al., 2003)(7)
65	13	1 103(5) (DP)	27 (5)	2.3(6)	Pacific Northwest (DePaola et al., 2002)
23	1	308 (MPN)	10 TDH+	3.2	Pacific Northwest (Kaysner et al., 1990b)

NOTES: (1) Pathogenic is defined as a Kanagawa-positive (KP+) or thermostable direct haemolysin-positive (TDH+). TDH is a toxin produced by *V. parahaemolyticus* that lyses red blood cells in Wagatsuma agar. (2) Number of isolates tested. Test methods: EB = enrichment broth followed by streaking on agar; DP = direct plating; MPN = most probable number. (3) Samples included oysters, water and sediment samples. (4) ND = not determined. (5) Isolates obtained from 36 oyster samples collected at or "near" maximum intertidal exposure. (6) Estimated mean percentage pathogenic from fitted Beta distribution. (7) This is a subset of the Cook et al. (2002) study.The studies representing different regions in the United States of America were analysed separately. The study by DePaola et al. (2002) was conducted in the Hood Canal area and represented the Pacific Northwest region. The study by Kaufman et al. (2003) was conducted on the Gulf Coast. It was assumed that the percentage pathogenic data from the Gulf Coast region could also be used to represent the Mid-Atlantic and Northeast Atlantic regions. This assumption was based on the data by Cook, Bowers and DePaola (2002) indicating no apparent difference in the percentage of *tdh+ V. parahaemolyticus* in oyster samples in the Gulf Coast, Mid-Atlantic, and Northeast Atlantic regions.

In Australia, Lewis et al. (2002) undertook a pilot study of prevalence of *V. parahaemolyticus* from oyster leases in NSW, South Australia and Tasmania. The organism was isolated from 16/20 (80%) of Sydney Rock oysters (*Crassostrea glomerata* syn. *commecialis*) from Wallis Lake in NSW; from 6/10 (60%) of Pacific oysters (*Crassostrea gigas*) in Tasmania; and 2/10 (20%) from South Australia. Previous reports had not mentioned the presence of pathogenic strains of *V. parahaemolyticus* in Australian waters or in marine products (see review by Desmarchelier, 2003). In the pilot study of Lewis et al. (2002), pathogenic *V. parahaemolyticus* were isolated from oysters from all three states, with between 10% and 20% of isolates being pathogenic strains (*tdh+*) at a concentration between 50/g (the limit of detection) and 350/g.

The study by Lewis et al. (2002), based on only 40 samples of oysters from three states, was not regarded as definitive in the quantitative sense and a longitudinal study over an annual cycle was undertaken. Madigan et al. (2007) investigated Pacific oysters in South Australia for presence of pathogenic vibrios. In 25 samples, each of twelve oysters, *V. parahaemolyticus* was isolated from four, of which three were *trh+* and none was *tdh+*. Interestingly, while sucrose-negative vibrios (a category that contains pathogenic strains) were relatively high (10^3–10^4/g) during warmer months, *V. parahaemolyticus* was isolated only after oyster samples were pre-enriched and molecular techniques were employed; when samples were enumerated the researchers found that pathogenic *V parahaemolyticus* was present at below the limit of detection (<10/g) in oyster meat. If more comprehensive Australian data become available, these could be used to replace United States of America surrogate data in the model.

The relationship between *V. parahaemolyticus* densities and water temperature based on the United States of America Pacific Northwest data was used for the Canadian, New Zealand and Japan simulations, as was the prevalence of *tdh+* *V. parahaemolyticus*.

I–2.7 Post-harvest

The Post-harvest Module describes the effects of typical industry practices—including transportation, handling and processing, as well as distribution, storage and retail—on *V. parahaemolyticus* numbers in oysters harvested from various locations and in different seasons. Factors considered as possible influences on the numbers of pathogenic *V. parahaemolyticus* at consumption include:

- ambient air temperatures at time of harvest;
- time from harvest until oysters are placed under refrigeration;
- time it takes oysters to cool once under refrigeration; and
- length of refrigeration time until consumption.

The purpose of modelling the Post-harvest Module is to simulate the effects of typical industry practices on the numbers of *V. parahaemolyticus* in oysters from harvest to consumption for various locations and seasons. The module can also be used to simulate the effect of intervention strategies. The inputs to the module are the regional and seasonal distributions of total and pathogenic *V. parahaemolyticus* at harvest. The outputs of the module are predicted distributions of the total and pathogenic numbers at time of consumption. The final steps to be addressed in the exposure assessment are the storage and retail conditions of the product, storage after retail, and finally preparation and consumption. Prior to consumption, temperature of storage is probably among the most critical factors to be considered. A diagrammatic representation of the parameters modelled in this module is presented in Figure I–6.

The principal assumption used to develop the relationships between numbers at harvest and at consumption is that the growth and survival of pathogenic *V. parahaemolyticus* is the same as total *V. parahaemolyticus*. Though no definitive studies of the growth characteristics of pathogenic *V. parahaemolyticus* are available, preliminary data suggest that there is little difference between growth characteristics of pathogenic *versus* non-pathogenic strains (DePaola, 1999). Furthermore, observation of the growth of total *V. parahaemolyticus* in oysters is limited to only one temperature (26°C). To bridge this data gap, a model of *V. parahaemolyticus* growth in broth developed by Miles et al. (1997) was used. The predictions of this model were adjusted to predict the growth rate in oysters, which is less than that of broth model systems, possibly due to the influence of competing microbiota.

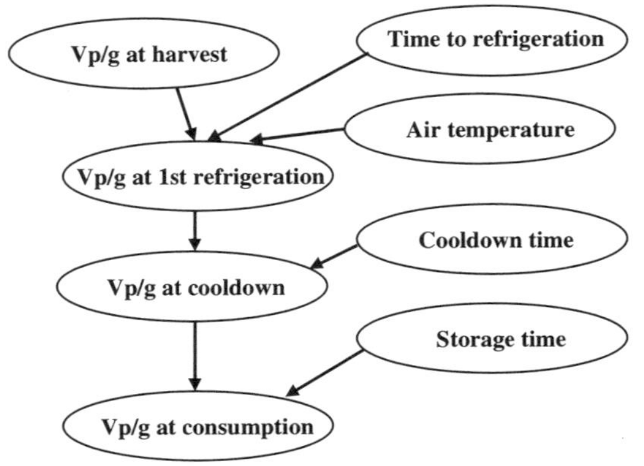

Figure I–6. Schematic depiction of the Post-harvest Module of the *V. parahaemolyticus* (Vp) risk assessment model.

I–2.7.1 Growth of *V. parahaemolyticus* from harvest to first refrigeration

The extent of growth that occurs from harvest until oysters are first placed under refrigeration is modelled by three factors:

- growth rate of *V. parahaemolyticus* as a function of temperature;
- temperature of oyster meat following harvest; and
- length of time product is unrefrigerated.

I–2.7.2 Distribution of ambient air temperature

Examination of water and air temperatures obtained from the NOAA/NBDC database in the United States of America showed a strong correlation between water and air temperature. This correlation has been incorporated into the risk simulation by modelling the distribution of the difference between water and air temperatures based on the normal distribution within any given region and season. These distributions are then used to predict the air temperature to which oysters would be subject, depending on the water temperature at the time of harvest.

In the process of simulating the distribution of total and pathogenic *V. parahaemolyticus* at harvest by the Monte Carlo method, the water temperature associated with any given outcome is retained. A corresponding air temperature is predicted by sampling from the appropriate distribution for the difference in air *versus* water temperature. This difference is then added to the water temperature to derive a corresponding air temperature. The distributions of difference in air *versus* water temperature were obtained by pooling the data available for each near-shore buoy across all available years. The mean and variance of these distributions are shown in Table I–14.

Air and water temperatures for Orongo Bay in New Zealand are presented in Table I–10, above, based on information from New Zealand Food Safety Authority (D.J. McCoubrey, pers. comm.).

I–2.7.3 Distribution of time oysters are left unrefrigerated

The distribution of the length of time that oysters are held unrefrigerated can be developed by using the distribution of length of working day, with the assumption that oysters are harvested uniformly from the start of the harvest up to one hour prior to conclusion of harvesting, when they are landed and placed in cold storage. Table I–15 shows the minimum, maximum and mean duration of oyster harvesting that we have projected for the different regions and seasons in the United States of America. In the risk simulation, BetaPERT distributions were used based on these parameters to simulate the variation in the duration of harvesting. A BetaPERT distribution is a translated and scaled Beta distribution with specified moments. It is commonly used for the purpose of simulating parameter variation within a defined range in Monte Carlo simulations.

Table I–14. Means and standard deviations of the distribution of the difference between recorded air and water temperatures (°C) at midday in the United States of America.

Region	Mean (standard deviation) distribution differences between air and water temperature			
	Winter (Jan–Mar)	**Spring** (Apr–Jun)	**Summer** (Jul–Sep)	**Autumn** (Oct–Dec)
Northeast Atlantic	-2.6 (5.0)	2.2 (3.2)	0.52 (2.7)	-3.2 (4.2)
Mid-Atlantic	-0.25 (4.0)	0.54 (2.9)	-1.4 (2.1)	-2.1 (3.1)
Gulf Coast	-1.07 (3.3)	-1.24 (1.63)	-1.66 (1.33)	-1.62 (3.3)
Pacific Northwest	-1.6 (1.8)	1.3 (1.3)	1.3 (1.5)	-0.8 (2.0)

SOURCE OF DATA: http://www.seaboard.nbdc.noaa.gov/Maps/Wrldmap.shtml

Table I–15. Duration (maximum, minimum, most likely) of oyster harvest operation for different regions and seasons in the United States of America.

Region	**Duration of harvest** (hours)			
	Winter (Jan–Mar)	**Spring** (Apr–Jun)	**Summer** (Jul–Sep)	**Autumn** (Oct–Dec)
Northeast Atlantic	11, 2, 8	11, 2, 8	11, 2, 8	11, 2, 8
Mid-Atlantic	11, 2, 8	11, 2, 8	11, 2, 8	11, 2, 8
Gulf Coast – LA	13, 7, 12	11, 5, 9	11, 5, 9	13, 7, 12
Gulf Coast – FL, AL, TX	11, 2, 8	10, 3, 7	10, 3, 7	10, 3, 7
Pacific Northwest	4, 1, 3	4, 1, 3	4, 1, 3	4, 1, 3

KEY TO STATES: AL = Alabama; FL = Florida; LA = Louisiana; TX = Texas.
SOURCES OF DATA: ISSC and FDA, 1997; Washington State Shellfish experts (pers. comm.) and Washington State Department of Health (Watkins, 2000).

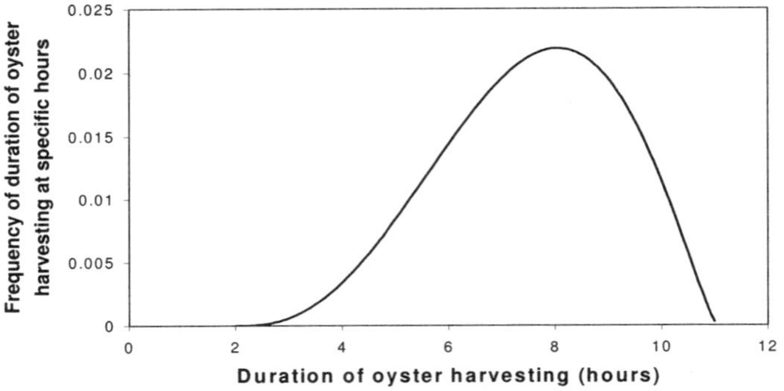

Figure I–7. BetaPERT probability density distribution for the duration of harvesting operations during the winter season (Mid-Atlantic, Northeast Atlantic, Gulf Coast, excluding Louisiana).

Figure I–7 shows the probability density of the BetaPERT distribution with a minimum of 2, maximum of 11 and mean of 8 hours.

For the United States of America Gulf Coast states, it was assumed that harvesting is done more quickly in the spring, summer and autumn due to the NSSP time-to-refrigeration requirements, and is generally longer in the winter. As indicated, harvesting of oysters was assumed to occur uniformly from start of harvest, up to one hour prior to the end of harvest operation. The distribution of the duration of time oysters were held unrefrigerated was simulated by first sampling from the distribution for duration of harvest operation and then sampling from a uniform distribution with a minimum of one hour and maximum corresponding to the randomly selected duration of harvest. Because they are harvested over the length of harvesting operations, the mean time that oysters remain unrefrigerated is almost always shorter than the maximum duration of harvesting.

Overall, the extent of growth occurring prior to time of first refrigeration (i.e. time at which oysters are first placed in refrigerated storage) was simulated by:

- sampling air temperature corresponding to the water temperature at harvest;
- sampling duration of harvest;
- sampling the length of time unrefrigerated given a particular duration of harvest; and
- calculating the extent of growth expected for the given duration unrefrigerated

In the other countries modelled in the present study, harvesting was very different from that in the United States of America (Table I–16), based on data being received via personal

Table I–16. Minimum, maximum and most likely duration of oyster harvest (length of harvesting operation in hours) for different countries.

Country	**Minimum harvest time** (hours)	**Most likely harvest time** (hours)	**Maximum harvest time** (hours)
Japan	0.5	1.5	6
Australia	2	4	6
New Zealand	1	3	5
Canada	0.1	0.75	1

communications from Hajime Toyofuku (Japan), Damian Ogburn (Australia), D.J. McCoubrey (New Zealand) and Klaus Schalle (Canada).

Canadian conditions reflect the legal requirement to begin chilling the shellfish no more than one hour after harvest. In Australasia, harvest times reflect the inter-tidal nature of the industry, while in Japan the most likely duration (1.5 hours) reflects the time from harvest to refrigeration.

I–2.7.4 Growth of *V. parahaemolyticus* during cooling

V. parahaemolyticus will continue to grow in oysters after they are placed under refrigeration until the temperature of the oyster tissues falls below a certain threshold (e.g. 10°C). The time it takes for oysters to cool once under refrigeration is assumed to vary according to the efficiency of the chilling medium, the quantity of oysters to be cooled and their arrangement in the cool room. Data on cooling rates of commercial oyster shellstock were not found. In the United States of America, preliminary GCSL experiments with a single in-shell oyster at 30°C, with a temperature probe inserted into its tissue, indicated a cooling rate of approximately 0.5°C/min when placed into a 3°C cooler (DePaola, 1999). However, 24 oysters in an uninsulated plastic container required approximately 7 hours to cool from 26°C to 3°C. These data suggest considerable uncertainty for cooling times after oysters are refrigerated, and it was concluded that a uniform distribution between 1 and 10 hours would be appropriate to describe the current state of knowledge.

As oysters cool down to storage temperatures it is reasonable to expect that the growth rate of *V. parahaemolyticus* slows with declining temperature. At the start of the cooling period, when oysters are first placed under refrigeration, the growth rate is still equal to the initial rate as determined by ambient air temperature. At the end of the cooling period, when oysters have reached storage temperatures, it was assumed that there is no further growth and that numbers will decline slowly thereafter (Gooch et al., 2002). Implicitly, this assumes that there is no appreciable temperature abuse after oysters have been placed in cold storage. As the rate at which oysters cool during cold storage is not known, it was assumed that during the period of cooling, the growth rate of *V. parahaemolyticus* falls uniformly to zero.

A discrete approximation of the extent of growth that may occur during cooling was simulated by first sampling from a discrete random uniform distribution between 1 and 10 hours (duration of cooling). The extent of growth during each hour of the cooling period was then estimated from the average growth rate during that hour. The average growth rates were dependent upon the growth rate of *V. parahaemolyticus* in oysters left unrefrigerated (i.e. as determined by the ambient air temperature for a given oyster lot) and the duration of cooling. The total excess growth was the sum of these values over the cooling period subject to the restriction that the maximum density of 6.0 $\log_{10}$ per gram could not be exceeded. These calculations are illustrated in Table I–17, where, for example, it takes k hours for a particular oyster lot to reach cooler temperature.

Table I–17. Discrete approximation of variation in the growth rate of *V. parahaemolyticus* during a cooling period of *k* hours.

Hour of the cooling period	Average growth rate ($\log_{10}$/hr) during the hour of cooling
1	$\frac{(k+1)-1}{k}\mu_m$
2	$\frac{(k+1)-2}{k}\mu_m$
3	$\frac{(k+1)-3}{k}\mu_m$
...	...
k	$\frac{(k+1)-k}{k}\mu_m$
k+1	0

Since the cooling time *k* is a random variable with a mean of 5.5 hours, the average extent of growth is $3.25*\mu_m$, where μ_m is the maximal growth rate determined by ambient air temperature at time of harvest. Thus, for an initial growth rate of 0.19 $\log_{10}$ per hour (i.e. at 26°C), the average growth occurring during cooling is approximately 0.6 $\log_{10}$, or almost two generations.

It is generally accepted that the Sydney Rock oyster (*Crassostrea glomerata* syn. *commercialis*) is extremely hardy and may be held out of water at ambient temperature for some days before release to the market; this may reflect the intertidal growing habit and consequent strengthening of the adductor muscle. For example, this species will keep for weeks at below 20°C without gaping and organoleptic quality is also maintained. It is normal handling practice to store this species at <20°C as soon as possible in the warmer months (November–April). Eyles et al. (1985a) also found that *Vibrio parahaemolyticus* grew poorly or not at all during storage of unopened Sydney Rock oysters at 15 and 30°C for 2 and 7 days. Although *V. parahaemolyticus* counts often increased at 30°C, counts above 10^4/g were not observed.

The total excess growth is the sum of the growth over the *k* hours:

$$\sum_{i=1}^{k}\mu_m * \frac{(k+1)-i}{k} = \mu_m * \left[(k+1) - \frac{1}{k}\sum_{i=1}^{k} i\right]$$

$$= \mu_m * \left[(k+1) - \frac{k+1}{2}\right]$$

$$= \mu_m * \frac{k+1}{2}$$

According to Australian regulations, oysters should be brought to no warmer than 10°C within 24 hours, though through a dispensation, Sydney Rock oysters may be held for up to 72 hours at 25°C before bringing to no warmer than 15°C. In practice, storage at such elevated temperatures almost never happens. If a rainfall event is forecast which might lead to closure, it is industry practice is to harvest and then hold the oysters at 15–20°C so that they can be released to the market during the next week or so. Once in the market place, oysters are chilled and kept refrigerated. Table I–18 presents data sum–marizing post-harvest temperature relations in Japan, New Zealand and Australia. These data were obtained via personal comm-unications from Hajime Toyofuku (Japan), Damian Ogburn (Australia), D.J. McCoubrey (New Zealand) and Klaus Schalle (Canada).

Table I–18. Minimum, most likely and maximum cooling times (hours) for oysters from different countries.

Country	Minimum cooling time	Most likely cooling time	Maximum cooling time
Japan	0.5	2	5
Australia	3	5	10
New Zealand	3	5	10

I–2.7.5 Change in *V. parahaemolyticus* population during cold storage

Gooch et al. (2002) showed that, in oysters, *V. parahaemolyticus* declined 0.003 $\log_{10}$ per hour when stored 14–17 days at 3°C. This die-off rate was assumed to be typical of all refrigerated oysters. Error may be introduced because commercial oysters are typically stored at higher temperatures (5–10°C). Die-off may have been overestimated because chill-stressed *V. parahaemolyticus* may not be recovered by the methods used in the study. One of the enumeration methods employed a repair step in a medium containing magnesium, which has been shown to increase recovery of chill-stressed cells. This method did not give higher *V. parahaemolyticus* counts after refrigeration than did the other methods that were used to calculate die-off. Therefore, the effect of chill-stress on die-off rate was assumed to be negligible.

Data from the FDA/ISSC retail study (see Cook et al. 2002) for the time between harvest and sample collection were assumed to be a reliable estimate for the refrigerated storage time to consumption (Cook, 1997a). Summary statistics on the storage time for samples obtained during the study are shown in Table I–19. A small degree of error may be introduced by assuming that these data are representative of storage time in so far as samples were generally collected on Monday or Tuesday and most servings are consumed in restaurants at weekends. Since this was a year-long nationwide survey, the mean of 7.7 days and range of 1 to 21 days was assumed to be representative of all seasons and regions. In the simulation, we used a BetaPERT distribution based on the overall mean, minimum, maximum and mode in order to obtain a smooth representation of the variation in the duration of storage time.

Table I–19. Summary United States of America statistics of the distribution of storage times (time under refrigeration in days) of oysters samples obtained during the ISSC/FDA retail study.

Storage Time	Consumed locally (within the same region as harvest)	Non-local (transported outside region of harvest)	Overall
Minimum	1	2	1
Maximum	20	21	21
Mean	6.3	9.9	7.7
Most likely	6	5	6

SOURCE OF DATA: FDA, 2005.

The predicted numbers of *V. parahaemolyticus* at time of consumption were therefore simulated by randomly sampling from the distribution of storage times and multiplying by a die-off rate of 0.003 $\log_{10}$ per hour. The resulting distribution was then subtracted from the predicted distribution of *V. parahaemolyticus* numbers in oysters initially reaching cooler (no-growth) temperatures.

Storage times for oysters in the Japanese, Australian and New Zealand systems are summarized in Table I–20, based on data received via personal communications from Hajime Toyofuku (Japan), Damian Ogburn (Australia) and D.J. McCoubrey (New Zealand); note that storage times for Canadian oysters were assumed to be identical with those of United States of America.

Table I–20. Minimum, maximum and most likely refrigerated storage time in days for oysters from different countries.

Country	Minimum refrigeration time (days)	Most likely refrigeration time (days)	Maximum refrigeration time (days)
Japan	0.5	6	12
Australia	1	6	10
New Zealand	1	2	5

I–2.8 Consumption

Food surveys and oyster landing statistics provide a basis for estimating the extent of exposure in the population. Distributions of ingested dose were developed by considering the probabilistic variation of number and meat weight of oysters in a serving, in addition to the expected variation of the numbers of pathogenic *V. parahaemolyticus* determined in the Harvest and Post-harvest Modules.

The primary data source used in estimating the number of oysters consumed was a consumer survey conducted by Degner and Petrone (1994). The data are a distribution of the self-reported numbers of oysters consumed per serving. As oysters are frequently sold in units of a dozen (twelve), the frequency distribution of numbers of oysters consumed spikes at numbers of half, one, two, three and four dozen. However, intermediate numbers of oysters consumed are also reported. In the simulation, the number of oysters consumed was modelled by selecting probabilistically the number of oysters consumed from the reported distribution.

The oyster meat weight was estimated using data from ISSC/FDA retail data (FDA/ISSC, 2000; DePaola, 2002). In this study, oyster weights were recorded from 339 samples of oysters collected from wholesale and retail locations. Each sample consisted of 4 to 16 (typically 12) oysters, and the weight was taken to be the weight of the oyster meat and mantel fluid, as both of these were likely to be consumed during an oyster meal. The average weight per oyster within each sample was obtained by dividing the weight of the sample by the number of oysters in the sample. The data were collected and a distribution of weights obtained. A log-normal distribution was fitted to the observed average oyster weight data. The geometric mean oyster weight was 15.2 g, with a geometric standard deviation of 1.4 g.

I–2.9 Mitigation strategies

This model can be used to demonstrate the effect of mitigation strategies. For example, the effects of three possible post-harvest mitigations can be evaluated in the Monte Carlo simulations:

- reduced time to refrigeration (i.e. rapid cooling);
- heat treatment; or
- freezing/cold storage

Mitigation through rapid cooling can be modelled by assuming that oysters will be cooled to no-growth temperatures immediately following harvest, by icing or otherwise cooling oyster shellstock aboard ship. In the simulation, it is assumed that the time unrefrigerated is zero (i.e. a degenerate distribution or constant). However, some growth is still projected to occur during cooling, as described earlier.

The effects of heat treatment and of freezing/cold storage can be evaluated by adjusting the simulated output of the baseline simulation (no mitigation), downward by factors of 4.5 $\log_{10}$ (the lowest level that caused a substantial reduction in illness after mild heat treatment) and 2 $\log_{10}$, respectively. Thus, random sequences of values for total and pathogenic numbers produced in the course of Monte Carlo simulation can be divided by 31 623 and 100, respectively. The implicit assumption is that the effect of treatment on $\log_{10}$ *V. parahaemolyticus* numbers is uniform, with no induced change in the variance of $\log_{10}$ numbers.

I–3. Hazard characterization

Dose-response relationships can be developed from epidemiological investigations of outbreaks and sporadic case series, human feeding trials or animal models of *V. parahaemolyticus* and related (surrogate) pathogens. In Japan, for example, human trials showed an increase in the number of illnesses with increasing numbers of pathogenic *V. parahaemolyticus*. Different dose-response models have been compared for the purpose of extrapolating risk of illness estimated on the basis of human feeding trials at high levels of exposure, to the lower levels of exposure associated with consumption of raw oysters (FDA, 2005). However, consideration of United States of America CDC estimates of annual illness suggests that the dose-response under conditions of population exposure is different from that observed in human volunteer studies. In other words, direct extrapolation of the dose response under conditions of exposure in the feeding trials is not supported by the epidemiological data. The human feeding trials were conducted under conditions of concurrent antacid administration. Due to possible food matrix effects of the oyster, dose response was shifted by 1 $\log_{10}$ from that estimated based on published clinical trials. This shift is derived from consideration of the CDC numbers of *V. parahaemolyticus* infections.

I–3.1 Description of the pathogen, host, and food matrix factors and how these influence the disease outcome

I–3.1.1 Characteristics of the pathogen

Infectivity, virulence/pathogenicity

Infection by *V. parahaemolyticus* is characterized by an acute gastroenteritis, usually within 4 to 30 hours from exposure. While most cases of *V. parahaemolyticus* infections are resolved without medical intervention, on rare occasions infection can lead to septicaemia and death. Not all strains of *V. parahaemolyticus* cause illness but most that do so are associated with the presence of a virulence factor, TDH, and possibly another virulence factor, the TDH-related haemolysin, TRH. The incidence of clinical infections can vary from year to year depending on the emergence of new strains.

Genetic factors, including antimicrobial resistance and virulence factors

The full number and role of virulence factors for *Vibrio* spp. is unknown and is the subject of continuing research. While a complete characterization of the virulence of each pathogenic *Vibrio* spp. strain and its associated virulence factors is not practical, several of the leading virulence factors have been characterized.

The virulence of *V. parahaemolyticus* appears to be largely attributable to TDH (Miyamoto et al., 1969). Strains of *V. parahaemolyticus* expressing this toxin lyse red blood cells on Wagatsuma agar and are also called Kanagawa Phenomenon positive (KP+); TDH+ and KP+ both indicate the presence of the toxin that is coded for by the gene *tdh+*. Typical KP-positive strains carry two *tdh* genes: *tdh*1 and *tdh*2 (Nishibuchi and Kaper, 1995). Other *tdh* genes, namely *tdh*3, *tdh*4 and *tdh*5, are found in weakly KP-positive strains. The *tdh+* allele is seldom found in environmental isolates of *V. parahaemolyticus*, but is frequently found in clinical isolates. KP-negative clinical isolates produce a TDH-related haemolysin, TRH, encoded by the genes *trh*1 or *trh*2 (Honda et al., 1988; Kinushita et al., 1992; Shirai et al., 1990). Some strains carry both *tdh* and *trh* genes, and the frequency of clinical cases occurring due to strains carrying only *trh* genes is much smaller than the frequency of cases due to strains carrying the *tdh* gene (Okuda et al., 1997).

I–3.1.2 Characteristics of the host

Immune status

The immune system of the host responds to *Vibrio* spp. infection to maintain health. The immunocompromised are at special risk for both infection and for more severe sequellae. In Japan, cases of *V. parahaemolyticus* bacteraemia have been reported among patients who were all immunosuppressed, especially with leukaemia and cirrhosis (Ng et al., 1999).

Age, sex and ethnic group

The vehicle of infection under consideration in this risk assessment for *V. parahaemolyticus* is raw oysters. The consumption patterns for raw oysters the United States of America have been estimated for age, sex and ethnic group (Desenclos et al., 1991; Timbo et al., 1995).

Anyone who consumes shellfish raw is "at risk" for infection by *V. parahaemolyticus*. FDA telephone surveys in the United States of America in 1993 and 1998 showed that consumption of raw shellfish is not uniformly distributed (Levy and Fein, 1999). A higher percentage of men consume raw oysters than women (16% *versus* 7%) and raw shellfish consumption is higher for those living along the coastline of the United States of America than for those living inland (22% *versus* 13%). The trend, as evident from the 1998 FDA survey, is toward lowered consumption of raw shellfish. This may be the result of education efforts by the FDA concerning risks associated with the consumption of raw or undercooked protein foods such as beef, chicken, eggs and shellfish. Raw shellfish consumption is highest among those with the highest education levels and the trend toward reduction in raw shellfish consumption over the last 5 years is smallest in this group.

In Japan, the proportion of oysters eaten raw is around 55% (Hiroshima prefectural government data[3]. In Australia and New Zealand, the proportion will be very much higher, with >90% of oysters thought to be eaten either raw or with minimal cooking (Nick Ruello, pers. comm.).

Japanese data (see Figure I–8) on age and sex distribution also suggests that every age group is susceptible to infection by *V. parahaemolyticus* (Anon., 1998b).

---

3. Data, from www.pref.hiroshima.lg.jp/www/contents/1170319316468/files/toukei.pdf [in Japanese]. This includes data from 2002 to 2009.

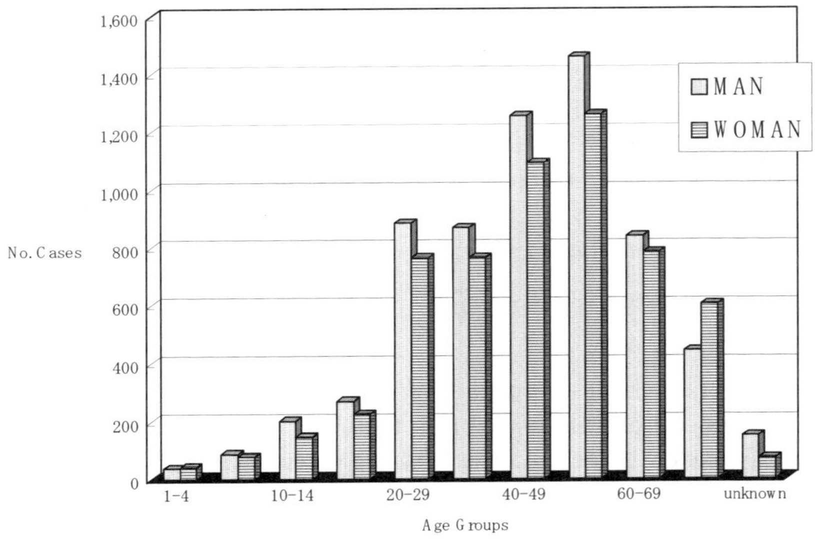

Figure I–8. Age distribution of patients infected with *V. parahaemolyticus* (1998 data; Japan).

Health behaviour

All *Vibrio* spp. are relatively susceptible to inactivation by cooking. Most of the risk associated with the relevant strains of *Vibrio* spp. in food comes from the consumption of raw seafood or from cross-contamination of other foods by raw seafood or contaminated water.

The time of year of consumption was considered in the risk assessment, as most infections occur during warm months. That is, a person consuming raw oysters in summer is at higher risk than a person consuming the same amount in winter. The location of harvest is also important. In the United States of America, for example, most landings of oysters occur in the Gulf of Mexico, particularly from off the coast of Louisiana, where the water temperature in the summer can reach the high twenties (°C), and with increases in salinity this results in environmental conditions that favour the survival and growth of *V. parahaemolyticus*.

Physiological status

While there are no known measures of physiological status relating to susceptibility to *V. parahaemolyticus* illness, analysis of epidemiological data indicate that pre-existing illnesses may predispose individuals with gastrointestinal illness to proceed to septicaemia (FDA, 2005).

In the United States of America, the average annual incidence of raw-oyster-associated illness from any *Vibrio* species among adults (>17 years of age) who consume oysters raw was estimated to be 10.1 per 1 000 000 (95% Confidence Interval (CI): 8.3–11.9). The annual incidence of fatal infections from any *Vibrio* species was estimated to be 1.6 per 1 000 000 adults who consume oysters (95% CI: 1.3–1.9). In two epidemiological studies (Hlady, 1997; Klontz, 1990), *V. parahaemo-*

lyticus accounted for 77/339 reported *Vibrio* infections (Table I–21). Of those 77 persons, 68 reported gastroenteritis and 9 had septicaemia; 29 persons were hospitalized for gastroenteritis, with no deaths reported, while 8 were hospitalized for septicaemia, of whom four died. Patients with septicaemia had underlying illnesses, including, but not limited to, cancer, liver disease, alcoholism and diabetes mellitus (Hlady, 1997; Klontz, 1990).

Hlady and Klontz (1996) reported that, of patients with infections, 25% had pre-existing liver disease or alcoholism. These included 75% of the septicaemia patients and 4% of the gastroenteritis patients. Of the remaining septicaemia patients, 9 reported having a history of at least one of the following: malignancy, renal disease, peptic ulcer disease, gastrointestinal surgery, diabetes, antacid medication or pernicious anaemia. Among the gastroenteritis patients, 74% had none of the above pre-existing medical conditions or had insufficient information to classify. Thus, while the prevalence of underlying illness was high in septicaemia patients, the majority of patients with raw-oyster-associated *Vibrio* gastroenteritis had no underlying conditions.

Table I–21. Clinical syndromes of raw-oyster-associated *Vibrio* infections in Florida, 1981–1994.

***Vibrio* species**	**Total cases**	**Gastroenteritis**	**Septicaemia**
V. vulnificus	95	13	82
V. parahaemolyticus	77	68	9
V. cholerae non-O1	74	8	66
V. hollisae	38	35	3
V. mimicus	29	29	0
V. fluvialis	19	19	0

SOURCE OF DATA: Hlady, 1997; Klontz, 1990.

During the first year of *Vibrio* surveillance in the United States of America (1989), *V. parahaemolyticus* accounted for 27/85 reported *Vibrio* illnesses characterized by gastroenteritis or septicaemia (Levine et al., 1993) and *V. parahaemolyticus* was the most prevalent of the *Vibrio* species reported. Twelve of the 27 persons with *V. parahaemolyticus* were known to have eaten raw oysters. One person had septicaemia while the remaining 26 persons had gastroenteritis. Oyster-associated infections occurred throughout the year, with a peak in October.

Table I–22. Underlying conditions in patients treated for *V. parahaemolyticus* infections (United States of America Gulf Coast states; 1988–1997).

Underlying medical condition	**Type of infection**	
	Gastroenteritis (263 cases)	**Septicaemia (20 cases)**
Diabetes	7%	17%
Peptic ulcer	6%	18%
Heart disease	6%	12%
Gastric surgery	4%	12%
Liver disease	—	63%
Alcoholism	3%	14%
Immunodeficiency	3%	18%
Haematological disease	2%	13%
Malignancy	2%	11%
Renal disease	1%	12%

Based upon the United States of America CDC surveillance data on *V. parahaemolyticus* from 1988 to 1997 in Alabama, Florida, Louisiana and

Texas (i.e. United States of America Gulf Coast states), the six most common underlying medical conditions associated with infection were diabetes, peptic ulcer, heart disease, gastric surgery, liver disease and immunodeficiency (Angulo and Evans, 1999). For gastroenteritis, 24% of respondents reported one or more of these six conditions, compared with 71% of respondents who had sepsis (Table I–22).

Genetic factors

There are no known human genetic factors that appear to be related to the susceptibility of individuals to *V. parahaemolyticus* illness.

I–3.1.3 Characteristics of the food matrix

Fat and salt content

Fat and salt content are probably not relevant in the determination of risk with respect to *Vibrio* spp. While the fat content of a matrix may be relevant with respect to the increase of effective dose of pathogens through protection of *Vibrio* spp. in micelles during gastric passage, there is insufficient evidence to model the degree of increased survival.

pH and water activity

Vibrio spp. appear to be relatively sensitive to both low pH and dehydration. Because of the nature of most foods associated with the unintended consumption of *Vibrio* spp., pH and water activity are probably not relevant in modelling survival of *Vibrio* spp. in raw seafood. However these parameters may be relevant in modelling the growth of *Vibrio* spp. in other foods as the result of cross-contamination.

I–3.2 Public health outcomes

I–3.2.1 Manifestations of disease

Gastroenteritis due to *V. parahaemolyticus* infection is usually a self-limiting illness of moderate severity and short duration (Barker, 1974; Barker and Gangarosa, 1974; Levine et al., 1993). However, severe cases requiring hospitalization have been reported. A summary of clinical features associated with *V. parahaemolyticus* gastroenteritis infection is presented in Table I–23. Symptoms include explosive watery diarrhoea, nausea, vomiting, abdominal cramps, and, less frequently, headache, fever and chills. On rare occasions, septicaemia, an illness characterized by fever or hypotension and the isolation of the micro-organism from the blood, can occur. In these cases, subsequent symptoms can include swollen, painful extremities with haemorrhagic bullae (Hlady, 1997; Klontz,

Table I–23. Clinical symptoms associated with gastroenteritis caused by *V. parahaemolyticus.*

Symptoms	Incidence of symptoms (%)	
	Median	**Range**
Diarrhoea	98	80–100
Abdominal cramps	82	68–100
Nausea	71	40–100
Vomiting	52	17–79
Headache	42	13–56
Fever	27	21–33
Chills	24	4–56

SOURCE OF DATA: Barker and Gangarosa, 1974; Levine et al., 1993.

1990). Duration of illness can range from 2 hours to 10 days (Barker and Gangarosa, 1974).

An outbreak is defined as the occurrence of two or more cases of a similar illness resulting from the ingestion of a common food. The incubation period ranges from 12 to 96 hours, with a median of approximately 15 to 24 hours. The number of raw oysters consumed ranges from 1 to 109 (median of 12). However, the duration of consumption is not known. The typical prevalence of symptoms for cases with gastroenteritis parallels those that were identified during the Pacific Northwest outbreak in the United States of America in 1997. These symptoms included diarrhoea (99%), abdominal cramps (88%), nausea (52%), vomiting (39%), fever (33%) and bloody diarrhoea (12%). Some outbreaks associated with *V. parahaemolyticus* that have occurred in the United States of America are listed in Table I–24.

Although *V. parahaemolyticus* outbreaks are less frequent in occurrence, sporadic cases are not infrequent, as further described below.

Several case reports have been published that outline clinical presentations and outcomes of patients with *V. parahaemolyticus*. One such describes a 35-year-old woman who sought medical attention for abdominal pain after she had consumed raw fish (Tamura et al., 1993). She presented with gastrointestinal symptoms, redness on lower extremities, fever, polyarthritis and weakness; *V. parahaemolyticus* was isolated in the stool culture and she was diagnosed as having reactive arthritis induced by *V. parahaemolyticus* infection.

Another clinical case report describes a 31-year-old female with a history of alcohol abuse, hepatitis C virus infection and cirrhosis (Hally et al., 1995). She presented with diarrhoea, weakness, leg pain and urine retention. The patient had ingested raw oysters and steamed shrimp 72 hours prior to admission. *V. parahaemolyticus* was isolated from blood samples. The patient suffered cardiac arrest and died six days after presentation.

Table I–24. Outbreaks associated with *V. parahaemolyticus* that have occurred in North America.

No. of persons ill	Location	Year	Food implicated	Reference
Not known	Maryland, USA	1971	Contaminated steamed crabs	Dadisman et al., 1972
40 outbreaks	15 states in the USA and Guam territories	1973–1998	Seafood or cross-contamination from raw or undercooked seafood	Daniels et al., 2000
209 persons	USA	1997	Oysters from USA (California, Oregon, Washington), and Canada (British Columbia)	CDC, 1998
6 culture-confirmed cases	North America	1981	Not known	Nolan et al., 1984
416	Texas. USA	1998	Raw oysters harvested from Galveston Bay, USA	Daniels et al., 2000
23 culture-confirmed cases	USA	1998 (May–Dec., with a peak in July-Aug)	Raw shellfish	CDC, 1999

A suspected case of a laboratory-associated infection was reported in 1972 (Sanyal et al., 1973). The day prior to the development of diarrhoeal disease, the laboratory worker had been handling *V. parahaemolyticus* strains for the first time. The illness was associated with severe upper abdominal pain, bloody stools, nausea and fever. Weakness and abdominal discomfort continued for two days beyond the onset of illness. No other source of *V. parahaemolyticus* could be identified and it was believed that the infection was caused by a relatively small inoculum (Sanyal et al., 1973).

A case series is a study of sporadic cases over a period. Sporadic cases of *V. parahaemolyticus* infections, while commonly reported by many states in the United States of America, are primarily reported by Gulf Coast states. Most *V. parahaemolyticus* infections present clinically as gastroenteritis, which has a low case fatality rate. Life threatening septicaemia can occur, especially in patients with underlying medical conditions. The case series has a range of infection throughout the year, with a peak in September to October. A case series of *Vibrio* infections related to raw oyster consumption was reported in Florida from 1981 to 1994 (Hlady, 1997).

I–3.3 Rationale for the biological end points modelled

Gastrointestinal illness is modelled as endpoint, corresponding to that measured in the human volunteer studies. Since most gastrointestinal illnesses are not reported, epidemiological data used in validating the model are scaled to account for under-reporting.

Dose-response relationships can be developed from epidemiological investigations of outbreaks and sporadic case series, human feeding trials or animal models of *V. parahaemolyticus* and related (surrogate) pathogens. In Japan, for example, human trials showed an increase in the number of illnesses with increasing numbers of pathogenic *V. parahaemolyticus*. Human volunteer studies are available to estimate the probability of illness given exposure based on dose-response investigations conducted by Sanyal and Sen (1974); Takikawa (1958); and Aiso (1963, cited in Aiso and Fujiwara, 1996).

Based on initial simulations, the number of illnesses estimated was higher than that predicted for the United States of America, a discrepancy thought to be due to the difference between the "administered dose" and the "effective dose". The administered dose is the number of viable *V. parahaemolyticus* cells given to the volunteer. The effective dose is the number of surviving cells that pass through the stomach and can begin the process of infection. The human volunteer studies used acid neutralizing solutions before administration of challenge doses in order to limit variation in human subject response due to variable reduction of challenge dose due to stomach acidity. This protocol is frequently used to minimize any difference between the administered dose and the effective dose. While this protocol helps to minimize variation of the effective dose, it complicates interpretation of the effective dose of pathogens in food. The exposure part of the model predicts an administered dose to a consumer. The model must determine the relationship between the administered doses of *V. parahaemolyticus* in food with the effective dose of *V. parahaemolyticus* that survives the stomach. Different foods may potentiate or diminish the effective dose. Data presented in the hazard characterization of *V. cholerae* demonstrate a dose response shift to the right between *V. cholerae* human volunteer studies where an acid-neutralizing solution was given before the challenge dose, compared with studies where a meat broth was given before the challenge dose. It is reasonable to expect a similar dose response shift to the right for *V. parahaemolyticus* administered with oysters compared with *V. parahaemolyticus* administered with acid-neutralizing solutions. For the present risk assessment, this dose shift was estimated at 27.1, applied to the beta factor of the Beta-Poisson dose functions (see Figure I–9).

The epidemiological data used in the estimate of the dose-response shift were based on an estimate of total number of illnesses, rather than on just illnesses reported to public health authorities. This estimate was obtained by multiplying the number of culture-confirmed *V. parahaemolyticus* illness cases recorded by United States of America public health authorities by a factor of 20 to correct for the estimated magnitude of under-reporting of illness in the United States of America. This is to say, in the United States of America, it is estimated that 20 cases of *V. parahaemolyticus* occur for every one reported case.

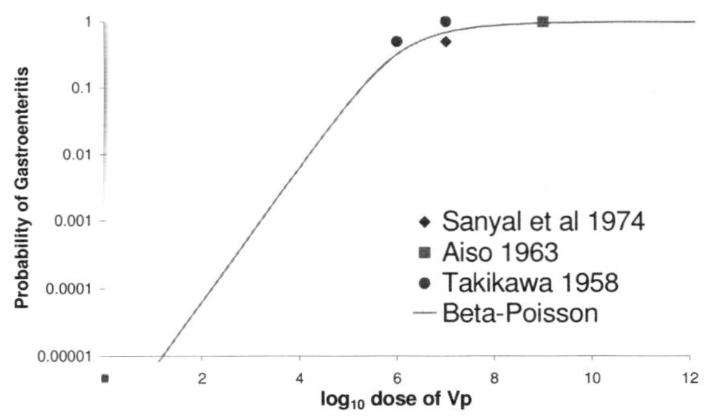

Figure I–9. Beta-Poisson dose-response curve for *Vibrio parahaemolyticus*.
SOURCE: from the FDA-VPRA (FDA, 2005).

Under-reporting of illness is a reality in all public health systems that rely primarily on passive reporting of illness. The reasons for under-reporting of illness are many. In some cases, the illness is mild enough that the individual does not consult a physician. In other cases, the health provider may not feel the illness is serious enough to warrant laboratory tests that would identify the causative organism. In still other cases, the wrong tests may be performed on the sample. Based on an epidemiological analysis that considers the severity of the disease and the likelihood that it can be properly identified, the CDC have estimated the factor of actual to culture-confirmed cases of *V. parahaemolyticus* infection in the United States of America (Mead et al., 1999)

The proper application of the model to harvesting areas outside the United States of America requires a comparable estimate of under-reporting of *V. parahaemolyticus* illnesses by the public health system of that area. Under-reporting factors for *V. parahaemolyticus* illnesses in the countries where the model is tested should be applied to predicted illnesses in order to compare the model's performance in predicting illness by its correlation with surveillance data. In the present study, in the absence of country-specific under-reporting factors, the United States of America factor of 20 was used in assessing model performance in predicting illness.

I–3.4 Dose-response relationship

I–3.4.1 Summary of available data

Probability of illness given exposure

Human volunteer studies are available to estimate the probability of illness given exposure. Sanyal and Sen (1974), Takikawa (1958) and Aiso (1963. cited in Aiso and Fujiwara, 1996) have conducted dose-response investigations.

Probability of sequellae given illness

In the United States of America, approximately 5% of culture-confirmed cases of *V. parahaemolyticus* progress to septicaemia (Angulo and Evans, 1999).

Probability of secondary and tertiary transmission

No reports were identified describing secondary or tertiary transmissions of illnesses caused by *V. parahaemolyticus*.

Probability of death given illness

Based on then United States of America statistics, around 20% of patients who are septicaemic with *V. parahaemolyticus* die (Angulo and Evans, 1999).

I–3.4.2 Dose-response model

Sources of data used

Figure I–9, taken from the FDA-VPRA (FDA, 2005), shows the maximum likelihood fit of the Beta-Poisson to the available feeding trial data. Due to the small number of subjects exposed during these studies there is considerable uncertainty about the best estimate of the dose response. Several statistical methods for characterizing the uncertainty of the dose-response parameters are available, including likelihood ratio-based confidence regions and bootstrapping techniques (parametric or non-parametric).

The results of a non-parametric bootstrap analysis of the *V. parahaemolyticus* data are shown in Table I–25. There is one entry in the table for each possible (realized or unrealized) outcome of the study. For each of these possible outcomes, a fit of the Beta-Poisson dose-response function was obtained by the maximum likelihood procedure and the resulting estimates of alpha and beta are shown (as well as the log of the ID_{50} of the fit obtained). The non-parametric bootstrap estimates of probability weight associated with each possible outcome and summary fit is denoted by "likelihood of re-sample". An analysis of deviance indicates that a good fit of the Beta-Poisson was obtained for most of the outcomes and those that are not adequately fitted by the Beta-Poisson model ($P<0.05$) are relatively unlikely.

For the purpose of the present assessment, the dose-response uncertainty was characterized using a non-parametric bootstrap procedure. If the true probabilities (risk) of illness at each study dose level were known, then the likelihood of the observed and alternative (possible but unrealized) outcomes would also be known. The likelihood of alternative outcomes (i.e. if the studies were to be replicated) could then be used to characterize the uncertainty of the parameter values for any dose-response function used to estimate (i.e. interpolate or extrapolate) the overall dose response. Given that the true probability of illness at each study dose level is unknown, bootstrap procedures substitute an estimate for these probabilities and, for the non-parametric bootstrap, that estimate is the observed frequency of illness.

The analysis of the dose-response data presented in Table I–25 can be used to characterize the effect of parameter uncertainty on simulation-based risk predictions by probability-weighted selection of the alpha and beta parameters for the dose-response segment of a Monte Carlo simulation run. For clarity, these alternative dose-response curves are not shown in Figure I–9.

Table I–25. Uncertainty of Beta-Poisson dose-response for *V. parahaemolyticus*: Maximum likelihood estimates (MLEs) of parameters and non-parametric probability (based on the observed data) associated with bootstrap re-samples.

ID	Possible resample					MLEs of parameters		Likelihood of re-sample	MLE of Log ID_{50}	Deviance of fit to resample	p-value of fit to resample
	x1	x2	x3	x4	x5	alpha	beta				
1*	0	0	0	0	4	1.47E+06	3.53E+14	0.00034	8.22	0.6450	0.8861
2*	0	0	0	1	4	1.26E+07	7.20E+14	0.00412	7.60	0.0857	0.9935
3*	0	0	0	2	4	636.53	1.65E+10	0.02058	7.26	0.1901	0.9792
4*	0	0	0	3	4	35.81	5.42E+08	0.05487	7.03	0.3262	0.9550
5*	0	0	0	4	4	20.84	1.99E+08	0.08230	6.83	0.5204	0.9144
6*	0	0	0	5	4	14.87	8.78E+07	0.06584	6.62	0.8557	0.8361
7*	0	0	0	6	4	10.58	2.99E+07	0.02195	6.31	2.2562	0.5210
8	0	0	1	0	4	3.89	2.28E+08	0.00069	7.65	7.4536	0.0588
9	0	0	1	0	4	1.31	2.93E+07	0.00823	7.31	4.4426	0.2175
10	0	0	1	0	4	0.52	3.61E+06	0.04115	7.00	2.9538	0.3988
11	0	0	1	0	4	0.47	1.50E+06	0.10974	6.70	1.7571	0.6243
12	0	0	1	0	4	0.60	1.31E+06	0.16461	6.46	0.9994	0.8014
13	0	0	1	0	4	1.00	1.80E+06	0.13169	6.26	0.6272	0.8902
14*	0	0	1	0	4	8.59	1.30E+07	0.04390	6.04	0.6242	0.8909
15	0	0	2	0	4	0.15	2.33E+05	0.00034	7.32	15.9553	0.0012
16	0	0	2	1	4	0.19	2.29E+05	0.00412	6.90	10.6999	0.0135
17	0	0	2	2	4	0.25	2.36E+05	0.02058	6.57	7.9684	0.0467
18	0	0	2	3	4	0.32	2.57E+05	0.05487	6.30	6.0785	0.1079
19	0	0	2	4	4	0.43	3.04E+05	0.08230	6.08	4.6970	0.1954
20	0	0	2	5	4	0.69	4.34E+05	0.06584	5.88	3.6564	0.3010
21*	0	0	2	6	4	6.92	4.49E+06	0.02195	5.68	2.3697	0.4993

NOTES: * unconverged estimates.

Assumptions

The primary assumptions are:

- Healthy volunteer responses to oral challenge are representative of the general population.
- Virulence of the pathogens or susceptibility of the host does not vary.
- Beta-Poisson dose-response model is reasonable for use in characterizing risk of illness when consuming *Vibrio* spp.

While these assumptions are not representative of what we believe is the nature of the human dose-response relationship, they form the basis of the first iteration of the hazard characterization. As more information becomes available, these assumptions can be revised to better reflect our understanding of dose-response relationships.

Goodness of fit of the distribution

The goodness of fit of the distributions for *V. parahaemolyticus* are uncharacterized as a family of dose-response parameters are used to represent the parameter uncertainty.

Uncertainty and variability in the estimates

This analysis incorporates both uncertainty and variability in the estimates. Since the dose-response estimates are based upon curves fitted to human volunteer data, there is uncertainty as to whether the parameters that give the best fit are the "true" parameters of the dose-response curve. To account for this uncertainty a Monte Carlo simulation model can be set up to select probabilistically from a group of plausible dose-response parameters obtained by statistical analysis of dose-response data. While it is not incorporated in the present assessment, it is reasonable to assume that there is variability in the virulence of the pathogens and in the susceptibility of the host. Further research is needed to provide data for assessing and modelling variability in pathogen virulence and in host susceptibility. A key uncertainty in this hazard characterization is the effect of the food matrix on the dose-response relationship. Both the FDA-VPRA (FDA, 2005) and evidence from the studies of Cash et al. (1974) have indicated that some food matrices may shift the dose-response curve to the right, indicating that a higher dose of the pathogen is required to cause illness. Future human volunteer studies may help to resolve this question and to provide data that will allow scaling factors to be applied to predict the risk for specific pathogens consumed with specific food matrices.

I–4. Risk characterization

As emphasized in the Introduction, a significant constraint was that data were available from only four countries, and then only to a limited extent. In the absence of country-specific data, United States of America data were used to enable the model to be run. The extent to which surrogate data were used is shown in Table I–26.

Table I–26. Data sources for the *V. parahaemolyticus* model.

	Australia	Canada	Japan	New Zealand
Water temperature	Local data	USA Pacific Northwest data	Local data	Local data
Relationship of water temperature and Vp levels	USA data	USA data	USA data	USA data
Time oysters are out of the water	Local harvest data	Local harvest data	Local harvest data	Local harvest data
Air temperature	Local data	Local data	Local data	Local data
Growth rate adjustment factor	USA data	USA data	USA data	USA data
Time oysters are refrigerated	Local data	Local data	Local data	Local data
Weight of oyster	USA data	USA data	USA data	USA data
Number of oysters consumed at one meal	USA data	USA data	USA data	USA data
Fraction of Vp that are *tdh*+ or *trh*+	USA data	USA data	USA data	USA data
Amount of oysters harvested from location	Local data	Local data	Local data	Local data
Multiplier for under-reporting	USA data	USA data	USA data	USA data
Fraction of oysters eaten raw	USA data	USA data	USA data	USA data

NOTES: Vp = *Vibrio parahaemolyticus*. *tdh*+ = Thermostable direct haemolysin-positive. *trh*+ = Thermostable-related haemolysin-positive.

I–4.1 Predicted illnesses from oysters consumed in countries of study

In Japan, production in Hiroshima Bay was taken as representative of that country's oyster production. Oyster production was given as a monthly estimate with one million yearly serves. The model predicts illnesses only in the October to December period. Note that oysters are not harvested during the summer months[4]. For example, oyster harvest in Hiroshima Bay extended from 29 October 2005 to 31

4. Oysters for raw consumption are not harvested in Japan in the summer months in the area where data were collected because microbiological criteria (MPN of coliform group in harvesting sea-water, the total plate count numbers, MPN of coliform groups and MPN of total *V. parahaemolyticus* in oyster) exceed the standards set by the Ministry of Health, Labour and Welfare (Ken Osaka, pers. comm.).

March 2006, when the water temperature at the 2-metre depth level ranged between 21° and 11°C.

In Australia, Wallis Lake was taken as typical of oyster harvesting areas of NSW, Australia. The model was adjusted to use a low water activity (A_w) value of 0.98 compared with the model's 0.985 to reflect Wallis Lake's higher salinity values.

For New Zealand, temperature data were made available for Orongo Bay as being representative of oyster production in New Zealand. Oyster production was given as a monthly estimate that totalled 600 000 servings. The model was adjusted to use a low A_w value of 0.98 compared with the model's 0.985 to reflect Orongo Bay's higher salinity values.

Canadian production was modelled as Pacific Northwest production, with changes in harvesting practice parameters. Canadian requirements are to bring oysters under temperature control with four hours of harvest when the air temperature is <15°C; at warmer temperatures oysters are to commence cooling within one hour of harvest.

Table I–27 presents estimates for predicted *V. parahaemolyticus* illness in the five countries modelled in the present risk assessment; note that warmer months in the Southern Hemisphere are the opposite of those in the Northern. With the exception of New Zealand, the model predicted higher levels of illness in the growing area studied than are recorded in each country's notifications (Table I–28).

In Japan, Hiroshima Bay production accounts for around 57% of the national output, equating to around 66 illnesses predicted annually. This may be compared with a figure of 13 notified cases of *V. parahaemolyticus* illness in the seven-year period (1998–2004).

In Australia, the model predicts more *V. parahaemolyticus* illnesses from this one growing area (Wallis Lake) than are reported for the entire country (National Enteric Pathogens Surveillance System, NEPSS). Wallis Lake accounts for 30% of NSW production and 27 predicted illnesses per annum translate to a total around 90 illnesses per annum. However, there have been only two cases involving *V. parahaemolyticus* and oyster consumption recorded for New South Wales in the past 18 years, and none in other states and territories. In 1992, Sydney Rock oysters were the suspected food vehicle for two cases, one of whom died (Kraa, 1995).

In the Orongo Bay area of New Zealand there are 22 marine farm leases, which account for around 15% of New Zealand total oyster production. A risk profile of *V. parahaemolyticus* in New Zealand seafoods was undertaken by Lake et al. (2003) and includes descriptions of all reported illnesses of *V. parahaemolyticus*. Although the organism is not notifiable in New Zealand a total of 32 sporadic cases were reported to the national identification database between 1998 and 2002. Many of the illnesses were associated with seafood imported privately from Pacific islands; there were no cases implicating oysters.

Table I–27. Predicted annual illnesses due to *V. parahaemolyticus* following consumption of oysters in Japan, Australia, New Zealand, Canada and United States of America.

	Jan-Mar	Apr-Jun	Jul-Sep	Oct-Dec
Hiroshima Bay, Japan	0	0	0	38
Wallis Lake, Australia	19	1	0	7
Orongo Bay, New Zealand	0	0	0	0
British Columbia, Canada	0	18	168	0
USA Gulf	10	698	1705	183
USA North Atlantic	0	3	14	2
USA Mid-Atlantic	0	4	7	4
USA Pacific NW	0	18	177	1

NOTE: The warmer months in the southern hemisphere are the opposite of those in the northern.

Table I–28. Predicted and reported annual illnesses due to *V. parahaemolyticus* following consumption of oysters in Japan, Australia, New Zealand and Canada.

	Annual cases predicted by model for the specified growing area	Proportion (%) of total production	Annual cases predicted by model for nation	Epidemiological evidence for *V. parahaemolyticus* illness from oyster consumption
Wallis Lake, Australia	27	30	91	2 cases in 18 years; two large outbreaks from other seafood sources
Orongo Bay, New Zealand	0	15	0	None during 1997–2002 from oysters; several outbreaks from other seafood sources
Hiroshima Bay, Japan	38	57.2	66	13 during 1998–2004
Br. Columbia, Canada	186	100	186	212 in decade 1997–2006 in British Columbia

NOTE: The warmer months in the southern hemisphere are the opposite of those in the northern.

Table I–29. Reported illnesses associated with *V. parahaemolyticus* in British Columbia, Canada.

	Total	Retail/restaurant	Self-harvest
1997	111	88	23
1998	15	9	6
1999	27	23	4
2000	8	2	6
2001	4	3	1
2002	17	12	5
2003	7	7	0
2004	6	6	0
2005	7	3	4
2006	19	8	11
Total	221	161	60

SOURCE: Personal communication from Liliana Rodriguez-Maynez, Canadian Food Inspection Agency.

Over the period 1997–2006 predicted cases for British Columbia in Canada (Table I–28), where the illness is notifiable, were higher than reported in health statistics for that province (Table I–29); notably, 60/221 (27%) of cases were from "non-commercial" oysters resulting from uncontrolled harvest. Because of the similarity of harvesting conditions with the United States of America Pacific Northwest, the models for Canada and the United States of America are similar. Canadian harvesting practices incorporate limits on how long harvested oysters may remain unrefrigerated after harvest, and this is incorporated into the Canadian model.

The United States of America predictions are presented for comparison with other harvesting areas and practices. The predictions are more fully developed and explained in FDA (2005). The estimates in Table I–30 are a comparison of mean risk per serving of *V. parahaemolyticus* illness for oyster meals from the respective harvesting areas. However, it must be emphasised that United States of America surrogate data underlie the risk estimates for the non-United States of America harvesting areas and comparison of between-country risk should be avoided.

Table I–30. Estimated risk of *V. parahaemolyticus* illness for oyster harvesting areas

		Risk of Vp Illness (rate of illness per consumption of meals)			
		Jan–Mar	**Apr–June**	**July–Sep**	**Oct–Dec**
USA Gulf	Mean	1.60E-06	1.45E-04	3.75E-04	3.08E-05
	5%'ile	4.13E-08	1.02E-05	2.86E-05	1.41E-06
	95%'ile	6.22E-06	4.68E-04	1.21E-03	1.09E-04
USA North Atlantic	Mean	1.11E-08	3.64E-06	1.78E-05	3.98E-07
	5%'ile	4.92E-10	8.35E-08	8.37E-07	1.25E-08
	95%'ile	3.47E-08	1.49E-05	6.86E-05	1.62E-06
USA Mid Atlantic	Mean	1.05E-08	3.11E-05	9.24E-05	2.21E-06
	5%'ile	4.93E-10	1.81E-06	4.86E-06	4.94E-08
	95%'ile	3.75E-08	0.000105	0.000331	1.02E-05
USA Pacific NW	Mean	1.24E-09	6.93E-06	7.71E-05	2.07E-07
	5%'ile	4.35E-11	1.33E-08	1.66E-06	1.87E-09
	95%'ile	4.85E-09	3.07E-05	0.000332	8.42E-07
Canada	Mean	7.46E-10	1.19E-07	1.08E-06	1.1E-08
	5%'ile	2.87E-11	2.01E-09	3.04E-08	4.81E-10
	95%'ile	2.75E-09	4.22E-07	4E-06	3.76E-08
Australia	Mean	6.1E-06	3.47E-07	5.97E-08	2.29E-06
	5%'ile	1.72E-07	1.9E-08	4.09E-09	7.59E-08
	95%'ile	2.29E-05	1.12E-06	2.04E-07	9.57E-08
New Zealand	Mean	3.25E-07	8.56E-08	1.4E-08	2.37E-07
	5%'ile	1.78E-08	5.24E-09	1.1E-09	1.34E-08
	95%'ile	1.12E-06	3.05E-07	4.57E-08	7.75E-07
Japan	Mean	3.65E-09	1.02E-07	1.00E-10	1.25E-04
	5%'ile	2.16E-10	4.96E-09	1.00E-10	5.02E-06
	95%'ile	1.3E-08	3.12E-07	1.00E-10	4.27E-04

I–4.2 Predicted surveillance results and under-reporting factors

Models that predict illnesses from diseases with symptoms that range continuously from mild to severe provide special challenges in reporting. While severe cases and outbreaks of these illnesses generally come to the attention of public health agencies, milder cases do not, for various reasons, and illness is under-reported. This presents challenges when, as in the present work, estimates of risk for different nations are analysed. Ideally, the total number of illnesses estimated (with uncertainties) will be comparable with epidemiological surveillance data. Realistically, for illnesses that present a range of symptoms, estimates will always be higher than reported numbers. The rate of under-reporting associated with the illness then becomes of pivotal importance.

In the present work, we used CDC's adjustment of 20 unreported *V. parahaemolyticus* illnesses for every reported illness. Unfortunately, we have no notion of the scale on which *V. parahaemolyticus* illness is unreported in Japan, Australia and New Zealand. What can be said is that each of these countries keeps statistics on illnesses caused by *V. parahaemolyticus*. In Japan, in particular, this pathogen is a major cause of food poisoning (Table I–31), though outbreaks and cases post-1999 seem to have diminished, perhaps due to the practice of harvesting oysters for raw consumption only when water temperatures are low.

Table I–31. Reported outbreaks and cases of *V. parahaemolyticus* illness in Japan

	Outbreaks	Cases
1999	667	9396
2000	422	3620
2001	307	3065
2002	229	2714
2003	108	1342
2004	205	2773
2005	113	2301

SOURCE: Ministry of Health, Labour and Welfare, Japan.

In British Columbia, cases (Table I–29) are based predominantly on laboratory confirmation, with a minority of cases based on an association of clinical symptoms and food history.

In Australia, there have been large outbreaks of *V. parahaemolyticus* in NSW, involving chilled, cooked shrimp imported from Indonesia. In 1990, more than 100 people were affected and one died, and, in 1992, two separate outbreaks involving >50 people occurred following consumption of cooked prawns (Kraa, 1995). Since these outbreaks, *V. parahaemolyticus* has been included in routine laboratory screening of cases of suspected food poisoning in NSW. Importantly in the specific case of oysters, following a large outbreak of Hepatitis A from Sydney Rock oysters from Wallis Lake in 1997 (Conaty et al., 2000), surveillance and management of the sector by the controlling authority (NSW Food Authority) was enhanced in the lead-up to the Sydney Olympic games.

The foregoing leads us to the conclusion that under-reporting of *V. parahaemolyticus* illness in Japan, Canada, New Zealand and Australia is probably no greater than in United States of America. Applying this ratio to Japan and British Columbia brings predicted and reported cases to the same order of magnitude, which might be considered an acceptable test for the model. In Australia, by contrast, there have been two reported illness in 18 years, compared with the model's prediction of around 1700 cases over the same period. Because the study of Lewis et al. (2002) provides only preliminary information on prevalence and concentration of pathogenic *V. parahaemolyticus* from Wallis Lake oysters, data from the US Pacific Northwest were used. These data, together with the water temperatures that obtain at Wallis Lake, account for the predicted illnesses. However, the study of Eyles et al. (1985a) shows clearly that *V. parahaemolyticus* does not grow in Sydney rock oyster stored at 30°C for seven days.

Within the constraints imposed by lack of local data and knowledge of under-reporting, the present assessment provides estimates of illness of *V. parahaemolyticus* from oysters that are of the same order of magnitude for three countries and, while for Australia the predicted:reported case ratio is greater, it might be that the oyster species studied is unique in its ability to prevent growth of *V. parahaemolyticus*.

I–5. Impact of establishing limits for the level of *V. parahaemolyticus* in oysters

At the 38th Session of CCFH the Committee reviewed the hygiene provisions in the Proposed Draft Standard for Live and Raw Bivalve Molluscs, which had been developed by the Codex Committee on Fish and Fishery Products. It was concluded that, while the standard includes various microbiological limits for bivalves, the basis for the proposed limits was not clearly established and there was no agreement as to what these limits should be. The report of the session indicated that the CCFH

> "..... request FAO and WHO to use the risk assessment on Vibrio parahaemolyticus in seafood, which they are developing to provide scientific guidance to the Codex Committee on Fish and Fishery Products, to follow up on the recommendations of the CCFH regarding the hygiene provisions in the Proposed Draft Standard for Live and Raw Bivalve Molluscs."

The risk management question proposed was:

> "Estimate the risk reduction from *V. parahaemolyticus* when the total number of *V. parahaemolyticus* or the number of pathogenic *V. parahaemolyticus* ranges from absence in 25 g to 1000 cfu or MPN per gram."

The estimation considered the impact of three different limits for *V. parahaemolyticus*: 100 cfu/g; 1000 cfu/g; and 10 000 cfu/g. These limits were considered to be applied when the products are cooled after harvesting, when the population of *V. parahaemolyticus* has stabilized, i.e. when the temperature becomes too low for further growth, but not so low that die-off occurs.

This estimation was carried out based on information from three countries: Australia, New Zealand and Japan. However, where the appropriate data were not available, surrogate data from the United States of America were used.

The estimation looked at the impact of all (100%) harvested oysters meeting a specified target limit, compared with the baseline distribution of *V. parahaemolyticus* for each of these countries, and it estimated the reduction in human illness together with the amount of product rejection that would occur if all product on the market were to meet the specified target. The estimates are presented in Tables I–32 and I–33.

Table I–32. Reduction in illness, based on meeting specified target numbers of *V. parahaemolyticus*, together with commensurate rejection of product for raw consumption.

Specified target	Reduction (%) in the number of predicted illnesses			Product (%) rejected to achieve these reductions in illness		
	Australia (summer)	New Zealand (summer)	Japan (autumn)	Australia (summer)	New Zealand (summer)	Japan (autumn)
100 cfu/g	99	96	99	67	53	16
1000 cfu/ g	87	66	97	21	10	5
10 000 cfu/g	52	20	90	2	1	1

Table I–33. *V. parahaemolyticus* illnesses predicted by meeting specified targets.

Specified target	Predicted number of illnesses per year		
	Australia (summer)	New Zealand (summer)	Japan (autumn)
100 cfu/g	Approx. 1 every 5 years	Approx. 1 every 10 years	Approx. 1 every 2 years
1 000 cfu/ g	1	1	1
10 000 cfu/g	5	3	4
No limit	17	4	38

I–5.1 Caveats

When considering these results a number of issues need to be taken into account:

- The baseline data (i.e. levels of *V. parahaemolyticus* in oysters) are different for each country and therefore each country's results must be considered individually because the reductions are relevant to their baseline only.
- In developing the risk assessment model, some surrogate data were used, as data on all the necessary parameters were not available for each country.
- These results relate to applying these limits after landing, at the cool-down stage and assume that it is possible to apply an appropriate sampling plan for testing at that point. It would also be possible to apply these criteria at another stage of the harvest-to-consumption chain, but further analysis would have to be undertaken to determine its impact.
- The targets considered here are for total numbers of *V. parahaemolyticus* (i.e. both pathogenic and non-pathogenic strains).
- These results (Tables I–32 and I–33) assume 100% compliance with the established limit. Note that risk assessments for other pathogen:commodity combinations have indicated that the level of compliance to an established limit is an important consideration in terms of the effectiveness of the limit as a risk reduction strategy.

I–6. Discussion and conclusions

The impact of the establishment of limits is something that can be evaluated using risk assessment, but the measures needed to achieve those limits are not considered.

The assessment was carried out for those periods of the year when *V. parahaemolyticus* numbers tend to be highest and the risk of illness is highest. For example, the assessment for Japan focused on autumn, as harvesting grounds in Japan are closed in the summer due to increased water temperature (regulation regarding the control of other pathogenic bacteria).

The variations in risk reduction and product rejection for each of the countries highlights the relationship between the specified target and the baseline levels of *V. parahaemolyticus* in oysters in a particular country, and emphasizes the fact that the establishment of international limits for *V. parahaemolyticus* may have greater impact on product rejection in some countries. For example, based on the model developed in this assessment, the establishment of a limit of 100 cfu/g implies rejection of 67% of Australian oysters for consumption as raw product, but would have much less impact on Japanese oysters.

Table I–32 indicates that the establishment of a limit can be an effective means to reduce risk to human health, provided there is compliance with that limit. However, the risk reduction to health comes at a price in terms of the amount of product that would potentially be rejected. In establishing such limits, a balance would have to be achieved between these two factors.

While the establishment of limits has merits in risk management, it must also be considered in the light of the actual reduction in illness. Table I–33 shows the predicted numbers of illness associated with the current level of *V. parahaemolyticus* in oysters, as well as the numbers predicted should each of the specified targets be enforced. As discussed earlier in this report, the levels of illness predicted on an annual basis by the risk assessment may differ from the actual number of reported cases, due in part to under-reporting of foodborne illness.

Appendix I–1 presents calculations that predict implications of non-compliance, without the need for further simulations.

I–6.1 Key outcomes of the risk assessment

The present risk assessment operated under the constraints described in the introduction and in Table I–26, whereby surrogate data were required to enable the model to be run. Some data gaps went to the very heart of the risk assessment, e.g. prevalence of *V. parahaemolyticus* in oysters, coupled with the proportion which were *tdh*+. Perhaps unsurprisingly, given the foregoing, estimates of illness for the four countries that were able to supply some required data did not always align with surveillance data for each country.

However, the present work has generated some positive outcomes:

- The model has been used to estimate illness from different oyster species grown under various regimes and regulatory management systems.

- The framework of the model is made available in this report for modification by risk assessors in other countries.
- Risk assessors will benefit from the accumulated knowledge presented in this report.
- Some countries, e.g. New Zealand and Australia, are investing in significant research studies on *V. parahaemolyticus* prevalence in oysters, and the proportion of pathogenic strains. This will allow these countries to refine predicted illnesses based on their own data.
- The model is an appropriate tool for testing mitigation strategies, both at harvest (such as reduced cooling times) and post-harvest by heating, freezing or high pressure treatment.

I–6.2 Gaps in the data

Deficiencies in the data available to conduct the present risk assessment were identified in order to suggest future research or further data gathering to reduce uncertainties. These areas are:

- Incidence or frequency of pathogenic *V. parahaemolyticus* in water and shellfish.
- Factors that affect incidence of pathogenic *V. parahaemolyticus* in the environment.
- Role of oyster physiology and immune status in affecting levels of *V. parahaemolyticus*. There is a need to correlate the number of *V. parahaemolyticus* with the percentage of oysters contaminated.
- More research on the potential virulence factors of pathogenic strains other than *tdh*, e.g. *trh*1 and *trh*2 enterotoxins. *V. parahaemolyticus* strains that do not produce TDH, TRH or urease have recently been found to induce fluid accumulation in suckling mice and diarrhoea in a ferret model after oral inoculation in a dose-dependent manner (Kothary et al., 2000). Correlation between clinical and environmental incidence of these strains has yet to be determined.
- Growth rate of *V. parahaemolyticus* within different oyster species at temperatures other than 26°C, including the issue of potential differences in the growth rate of pathogenic strains *versus* total *V. parahaemolyticus* populations.
- Rates of hydraulic flushing (water turnover) in shellfish harvest areas based on levels of freshwater flows, tidal changes, winds and depth of harvesting area, and how these factors may influence pathogenic *V. parahaemolyticus* numbers.
- Consumer handling of oysters.
- Improved global public health surveillance of *V. parahaemolyticus* to identify new epidemic strains as they emerge.
- Knowledge of reporting systems in each country of study; this is of specific importance when under-reporting is taken into account to compare estimated with reported illnesses.

I–7. Current and future issues influencing risk assessment and management of *V. parahaemolyticus* in oysters

Risk assessments are typically conducted over a significant period, and the present assessment is no exception, having its genesis in 2001. It is axiomatic that, given the pace of scientific progress, any risk assessment process will always lag somewhat. In an attempt to accommodate recent advances and events surrounding *V. parahaemolyticus*, the present section has been written. As such, it provides an up-to-date scientific context on the key premises underpinning the assessment, and speculates on how this risk assessment model might be used and developed by risk managers as growing waters become progressively warmer and *V. parahaemolyticus* extends its latitudinal reach.

This risk assessment predicts the risk of *V. parahaemolyticus* illness from consumption of raw oysters harvested from selected Pacific Rim countries based on the framework previously developed to predict risk of illness in the United States of America (FDA, 2005). It was constructed around four key premises:

- the risk of infection is proportional to the exposure to pathogenic strains defined as those possessing the *tdh* gene and that all strains possessing *tdh* are equally pathogenic;
- levels of pathogenic *V. parahaemolyticus* can be reliably estimated from total *V. parahaemolyticus* levels, which are easier to measure;
- *V. parahaemolyticus* levels in oysters at harvest can be predicted based on water temperature; and
- total and pathogenic *V. parahaemolyticus* grow and survive equally during post-harvest handling and processing.

In the FDA-VPRA (FDA, 2005), a market study of raw oysters that was not used to estimate model parameters was used to validate the exposure assessment. The reliability of the FDA-VPRA's ability to predict exposure to total *V. parahaemolyticus* levels at or near the point of consumption was demonstrated for each season. However, regional attribution of illnesses as determined by the risk characterization was not consistent with estimates based solely on reported illnesses for which trace-back information was available. While some of these differences may be due to regional biases in reporting of illness or trace back, they also suggest that some premises, or assumptions, underpinning the FDA-VPRA might be suspect. Considerable new information on these key premises has become available since the FDA and FAO/WHO VPRAs were initiated. These new findings are discussed below, and may provide guidance for future efforts to refine the current versions of the FDA-VPRA and to implement these models for risk management.

The premise that *V. parahaemolyticus* risk is proportional to exposure to levels of pathogenic *V. parahaemolyticus* continues to be supported by epidemiological data, as ca. 90% of clinical isolates reported to CDC from 2001 to 2004 possessed the *tdh* gene.

However, an increasing proportion of clinical isolates possessed neither the *tdh* nor *trh* genes, and these were associated with the most severe cases, requiring hospitalization. It is unknown whether these isolates are related to each other, or the types of virulence attributes they possess. One of these strains has been selected for whole genome sequencing; mining of these data may reveal potential virulence determinants. The assumption that all *tdh*+ strains are equally virulent was based more on the lack of data to the contrary rather than targeted studies supporting this hypothesis. Epidemiological studies of the 2004 Alaskan cruise ship outbreak have produced perhaps the strongest challenge to this assumption. The outbreak was associated with serotype O6:K18, which caused a 30% attack rate among passengers consuming one to six oysters during three cruises over a two-week period in July 2004 (McLaughlin et al., 2005). Oysters collected from the implicated farm during this period typically contained less than 10 MPN/g of the outbreak strain. The FDA-VPRA dose-response curve indicates that a dose ca. 10 000-fold greater would be necessary to cause a 30% attack rate. The O6:K18 isolates from the Alaskan outbreak were indistinguishable by PFGE from those isolated in sporadic cases from Pacific Coast states over the previous decade. It is unclear whether this strain can infect at lower doses than the strains used in previous human volunteer studies. Another possible explanation is that the liquid doses used in human volunteer studies may underestimate the infectivity of *V. parahaemolyticus* relative to consumption of raw oysters that have naturally accumulated *V. parahaemolyticus*.

New studies employing real-time PCR analysis of oyster:alkaline peptone water enrichments in an MPN format are addressing the second key premise of this FDA-VPRA, namely that levels of pathogenic *V. parahaemolyticus* can be reliably estimated from total *V. parahaemolyticus* levels (Miwa et al., 2003; Nordstrom et al., 2007). This PCR-MPN approach permits simultaneous examination of thousands of *V. parahaemolyticus* cells for *tdh* and *trh* genes from each MPN tube, and is much more sensitive than cultural methods that require colony isolation and which typically only examine a few isolates for these genes. However, the recent detection of the *trh* gene in other *Vibrio* spp., such as *V. alginolyticus* and *V. harveyi* (Gonzalez-Escalona et al., 2006; Masini et al., 2007), complicates the interpretation of such assays. Even among *tdh*+ strains, expression of the gene varies depending on the type of allele present. Expression of TDH is much higher in strains possessing the *tdh*2 gene compared with other variants of *tdh* (Nishibuchi and Kaper, 1995; Nakaguchi and Nishibuchi, 2005). Thus, strains with other variants of the *tdh* gene may be less pathogenic compared with strains possessing the *tdh*2 gene. These studies are beginning to show that the relationship between total and pathogenic *V. parahaemolyticus* is far more variable than assumed in the FDA-VPRA. For instance, in the summers of 2004 and 2005, ecological studies of *V. parahaemolyticus* in Alaska indicated that pathogenic *V. parahaemolyticus* accounted for 74% and 30%, respectively, of the total *V. parahaemolyticus* population of oysters. Employing the same methods in Alabama and Mississippi in the summer of 2004, pathogenic *V. parahaemolyticus* accounted for 0.001% and 0.05% of the *V. parahaemolyticus* population, respectively (DePaola et al., 2007). Thus, consumption of Alaskan oysters containing 10 total *V. parahaemolyticus* per gram may be more risky than consuming Alabama oysters with levels greater than 100 000 total *V. parahaemolyticus* per gram, assuming that the *tdh*+ strains from Alaska and Alabama are equally virulent.

The third key premise of the FDA-VPRA, namely that *V. parahaemolyticus* levels in oysters at harvest can be predicted by water temperature, still appears to be on fairly solid ground for total *V. parahaemolyticus*. When water temperatures are below 15°C, *V. parahaemolyticus* levels are generally below one per gram, and outbreaks do not occur. However, data is beginning to accumulate indicating that the ratio of pathogenic to total *V. parahaemolyticus* may be affected by water temperature, with higher prevalence of pathogenic *V. parahaemolyticus* occurring at lower water temperatures. Regionally, the proportion of *V. parahaemolyticus* population in United States of America oysters possessing the *tdh* gene appears to be increasingly moving north from the Gulf Coast into the Pacific Northwest and Alaska. In a study of Alabama oysters, the proportion of pathogenic *V. parahaemolyticus* was found to be ca. 10-fold greater when the water temperature was <20°C compared with >20°C (DePaola et al., 2003). Water temperature alone has been estimated to account for approximately 50% of the annual variability in total *V. parahaemolyticus* levels in oysters. Recent studies suggest that salinity and turbidity may also influence *V. parahaemolyticus* levels in oysters, and incorporation of these parameters into the *V. parahaemolyticus* model may further reduce uncertainty of model predictions. Water temperature, salinity and turbidity can all be measured daily over the entire planet by satellite imagery. The feasibility of integrating risk assessment models into remote sensing data for prediction of *V. parahaemolyticus* levels in real time was demonstrated in a recent scientific publication. This approach was suggested as a management tool for *V. vulnificus* in oysters at the 2006 FAO/WHO Kiel Conference for developing practical risk management applications from quantitative microbiological risk assessments (FAO/WHO, 2006).

There has been limited research addressing the fourth and final key premise, that total and pathogenic *V. parahaemolyticus* grow and survive equally during post-harvest handling and processing. From the point of controlling oyster-borne *V. parahaemolyticus* illnesses, this is probably the most important premise, since harvest is the point when humans take control of *V. parahaemolyticus* levels from nature. FDA submitted a proposal to the 2007 ISSC in Albuquerque, United States of America, to limit *V. parahaemolyticus* growth in oysters post-harvest to half of a log. The FDA-VPRA predicts that achieving an average of half a log reduction in *V. parahaemolyticus* levels in oysters would result in ca. 70% reduction in *V. parahaemolyticus* illnesses associated with consumption of United States of America oysters. There is a major study underway at the University of Maryland-Eastern Shore addressing this assumption, but the completion of this study was not anticipated until 2008 (Parveen et al., 2007). Preliminary results from storage of Alaskan oysters at ambient air temperature (15–20°C) for up to 5 days indicated that growth rates of total and pathogenic *V. parahaemolyticus* were nearly identical and were in agreement with growth rates predicted by the FDA-VPRA at these temperatures. There is some information that the O3:K6 pandemic strains are more pressure resistant than other *V. parahaemolyticus* strains (Cook, 2003). However, in an FDA/ISSC survey of post-harvest treated oysters in the United States of America (mild pasteurization, freezing and high hydrostatic pressure), approximately 40 *V. parahaemolyticus* strains that survived these treatments were isolated, primarily from 25-g enrichments (unpublished data). A single *tdh+* isolate was observed among these strains, suggesting that *tdh+ V. parahaemolyticus* responds to these post-harvest treatments in a similar manner to non-pathogenic *V. parahaemolyticus*.

In the past decade, trends in *V. parahaemolyticus* illnesses associated with molluscan shellfish have taken unexpected and unprecedented turns. Pandemic *V. parahaemolyticus* has now spread to at least five continents, and has caused repeated annual outbreaks in Chile, where *V. parahaemolyticus* was not even considered a problem when this risk assessment began in 2001. One of the outbreaks in Chile was of epidemic proportions, causing over 10 000 cases, most of which were associated with consumption of molluscan shellfish (Fuenzalida et al., 2006).

More of the scientific community is coming to the conclusion that ballast discharge is a major mechanism for global spread of pandemic *V. parahaemolyticus*. While global trade continues to grow, there are few if any control measures implemented to protect molluscan shellfish from contamination. Vibrios appear to be uniquely adapted for dissemination by ballast discharge, and it is likely that new strains will emerge and, in the absence of controls, be spread globally.

In the past decade, the warmest temperatures on record have occurred in many parts of the world that are major producers of molluscan shellfish. Warmer temperatures may in part be responsible for the CDC-reported increase of *Vibrio* infections in United States of America, while illnesses of other major foodborne pathogens are declining (CDC, 2007). Since 2000, November has become the leading month for *V. vulnificus* infections associated with oyster consumption in the United States of America (Mark Glatzer, FDA, pers. comm.). Warmer temperatures also appear to be the cause of *V. parahaemolyticus* expanding its geographical range into areas such as Alaska (McLaughlin et al., 2005), Europe (Martinez-Urtaza et al., 2005) and Chile (Gonzales-Escalona et al., 2005). Alaska is 1000 km further north than any previous reported *V. parahaemolyticus* illnesses. The unprecedented high prevalence of pathogenic strains among the *V. parahaemolyticus* population of Alaskan oysters, and perhaps higher virulence compared to other pathogenic strains, was surprising and remains unexplained.

The Alaskan oyster industry could not wait for elucidation of *V. parahaemolyticus* ecology in the Prince William Sound, as their industry was fragile and at risk of collapsing if outbreaks continued. Instead, they relied heavily on the FDA-VPRA and, in 2005, voluntarily adopted the control of sinking oyster cages below the thermocline where water temperatures were <10°C. Under environmental conditions similar to those of 2004, when 62 *V. parahaemolyticus* cases were reported, there was only a single case reported in 2005 after implementing this control (Manny Soares, Alaska Department of Conservation, pers. comm.). In response to the 1997 and 1998 *V. parahaemolyticus* outbreaks in the Pacific Northwest, the Canadian government implemented strict time × temperature controls for oysters harvested in British Columbia, and no major outbreaks have since occurred. Washington State, United

States of America, which has similar environmental conditions to British Columbia, but less stringent time and temperature controls, experienced a major outbreak of 177 cases in 2006, most of which were associated with consumption of raw oysters (CDC, 2006). In most areas of the world, it may take decades before a system is available to reliably predict levels of pathogenic *V. parahaemolyticus* in the environment. As new and more powerful methods for measuring pathogenic *V. parahaemolyticus*, such as real-time PCR, obtain more widespread acceptance and application, areas and times of high risk should become more apparent. At the same time, the effectiveness of pro-active post-harvest controls based on the existing FDA-VPRA have already been demonstrated in the real world, and this approach currently offers the best available opportunity to control *V. parahaemolyticus* illnesses associated with consumption of raw oysters and other raw molluscan shellfish.

I–8. References for Part I – Oysters

Abbott, S., Powers, C., Kaysner, C.A., Takeda, Y., Ishibashi, M., Joseph, S.W. & Janda, J.M. 1989. Emergence of a restricted bioserovar of *Vibrio parahaemolyticus* as the predominant cause of *Vibrio*-associated gastroenteritis on the West Coast of the United States and Mexico. *Journal of Clinical Microbiology*, 27: 2891–2893.

Abd, H, Weintraub, A. & Sandstrom, G. 2005. Intracellular survival and replication of *Vibrio cholerae* O139 in aquatic free-living amoebae. *Environmental Microbiology*, 7(7): 1003–1008.

Aiso, K. & Fijuwara, K. 1996. Feeding tests of the pathogenic halophilic bacteria. *Annual Research Report, Institute of Food Microbiology, Chiba University*, 15: 34–38.

Amako, K., Shimodori, S., Imoto, S., Miake, T.S. & Umeda, A. 1987. Effects of chitin and its soluble derivatives on survival of *Vibrio cholerae* O1 at low temperature. *Applied and Environmental Microbiology*, 53: 603–605.

Angulo F. & Evans M. 1999. *Vibrio parahaemolyticus* epidemiological data. Personal communication. United States Centers for Disease Control and Prevention.

Anon[ymous]. 1996. *Boletin Epidemiologico Semanal*, 4(46): 401.

Anon. 1998a. *Boletin Epidemiologico Semanal*, 6(31): 306.

Anon. 1998b. National Nutrition Survey, Japan. Ministry of Health, Labour and Welfare, Japan.

Anon. 1999a. *Vibrio parahaemolyticus*, Japan 1996–1998. *Infectious Agents Surveillance Report*, 20(7): 1–2.

Anon. 1999b. *Vibrio parahaemolyticus*, Taiwan: Background. PROMED-digest, 28 May 1999.

Baffone, W., Pianetti, A., Bruscolini, F., Barbieri, E. & Citterio, B. 2000. Occurrence and expression of virulence-related properties of *Vibrio* species isolated from widely consumed seafood products. *International Journal of Food Microbiology*, 54: 9–18.

Barker, J. & Brown, M.R.W. 1994. Trojan horses of the microbial world: protozoa and the survival of the bacterial pathogens in the environment. *Microbiology*, 140: 1253–1259.

Barker, W.H. 1974. *Vibrio parahaemolyticus* outbreaks in the United States. pp. 47–52, *In:* T. Fujino, G. Sakaguchi, R. Sakazaki and Y. Takeda (editors). International Symposium on *Vibrio parahaemolyticus*. Saikon Publishing Company, Tokyo, Japan.

Barker, W.H. & Gangarosa, E.J. 1974. Food poisoning due to *Vibrio parahaemolyticus*. *Annual Review of Medicine*, 25: 75–81.

Baross, J.A., Liston, J. & Morita, R.Y. 1978. Incidence of *Vibrio parahaemolyticus* bacteriophages and other *Vibrio* bacteriophages in marine samples. *Applied and Environmental Microbiology,* 36: 492–499.

Bates, T.C., Tolker-Nielsen, T., Molin, S. & Oliver, J.D. 2000. The viable but nonculturable state in *Vibrio parahaemolyticus*. Abstracts of the 100th General Meeting of the American Society for Microbiology. Abstract No. I-87:400.

Bean, N.H., Maloney, E.K., Potter, M.E., Korazemo, P., Ray, B., Taylor, J.P., Seigler, S. & Snowden, J. 1998. Crayfish: a newly recognized vehicle for *Vibrio* infections. *Epidemiology and Infection,* 121: 269–273.

Bouchriti, N., El Marrakchi, A. & Goyal, S. 1995. Occurrence of marine vibrios in Moroccan coastal waters and shellfish. *Microbiologie Aliments Nutrit.* 13: 381–387.

Buchanan, R.L., Whiting, R.C. & Damert W.C. 1997. When is simple good enough: A comparison of the Gompertz, Baranyi, and three phase linear models for fitting bacterial growth curves. *Food Microbiology,* 14: 313–326.

Cash, R.A., Music, S.I., Libonati, J.P., Snyder, M.J., Wenzel, R.P. & Hornick, R.B. 1974. Response of man to infection with *Vibrio cholerae*. I. Clinical, serologic, and bacteriologic responses to a known inoculum. *Journal of Infectious Diseases,* 129: 45–52.

Cawley, P. & Norris, D. 1973. Isolation of *Vibrio parahaemolyticus* in New Zealand. *New Zealand Medical Journal,* 78: 107.

CDC [Centres for Disease Control and Prevention]. 1998. Outbreak of *Vibrio parahaemolyticus* infections associated with eating raw oysters – Pacific Northwest, 1997. *Morbidity and Mortality Weekly Report,* 47: 457–462.

CDC. 1999. Outbreak of *Vibrio parahaemolyticus* infection associated with eating raw oysters and clams harvested from Long Island Sound – Connecticut, New Jersey and New York, 1998. *Morbidity and Mortality Weekly Report,* 48: 48–51.

CDC. 2006. *Vibrio parahaemolyticus* infections associated with consumption of raw shellfish – three states, 2006. *Morbidity and Mortality Weekly Report,* 55: 854–856.

CDC. 2007. Preliminary FoodNet data on the incidence of infection with pathogens transmitted commonly through food – ten states, 2006. *Morbidity and Mortality Weekly Report,* 56: 336–339.

Chan, K.Y., Woo, M.L., Lam, L.Y. & French, G.L 1989. *Vibrio parahaemolyticus* and other halophilic vibrios associated with seafood in Hong Kong. *Journal of Applied Bacteriology,* 66: 57–64.

Colwell, R.R., West, P.A., Maneval, D., Remmers, E.F., Elliot, E.L. & Carlson, N.E. 1984. Ecology of pathogenic vibrios in Chesapeake Bay. pp. 367–387, *In:* R.R. Colwell (editor). *Vibrios in the Environment.* A Wiley-Interscience publication, John Wiley and Sons, New York, USA.

Colwell, R.R., Brayton, P.R., Grimes, D.S., Roszak, D.B., Huq, S.A. & Palmer, L.M. 1985. Viable but non-culturable *Vibrio cholerae* and related pathogens in the environment: implications for release of genetically engineered micro-organisms. *Bio/Tech.* 3: 817–820.

Comeau, A.M., Buenaventura, W. & Suttle, C.A. 2005. A persistent, productive, and seasonally dynamic Vibriophage population within Pacific oysters (*Crassostrea gigas*). *Applied and Environmental Microbiology,* 71: 5324–5331.

Conaty, S., Bird, P., Bell G., Kraa, E., Grohmann, G. & McAnulty, J. 2000. Hepatitis A in New South Wales, Australia, from consumption of oysters: the first reported outbreak. *Epidemiology and Infection,* 124: 121–130.

Cook, D.W. 1994. Effect of time and temperature on multiplication of *Vibrio vulnificus* in postharvest Gulf coast shellstock oysters. *Applied and Environmental Microbiology,* 60: 3483–3484.

Cook, D.W. 1997a. Refrigeration of oyster shellstock: conditions which minimize the outgrowth of *Vibrio vulnificus*. *Journal of Food Protection,* 60: 349–352.

Cook, D.W. 2002. Comparison of growth rate of pathogenic and non-pathogenic strains of *Vibrio parahaemolyticus*. Personal communication (unpublished data).

Cook, D.W. 2003. Sensitivity of *Vibrio* species in phosphate-buffered saline and in oysters to high pressure processing. *Journal of Food Protection,* 66: 2276–2282.

Cook, D.W. & Ruple, A.D. 1989. Indicator bacteria and *Vibrionaceae* multiplication in post-harvest shellstock oysters. *Journal of Food Protection,* 52: 343–349.

Cook, D.W. & Ruple, A.D. 1992. Cold storage and mild heat treatment as processing aids to reduce the numbers of *Vibrio vulnificus* in raw oysters. *Journal of Food Protection,* 55: 985–989.

Cook, D.W., O'Leary, P., Hunsucker, J.C., Sloan, E.M., Bowers, J.C., Blodgett, R.J. & DePaola, A. 2002. *Vibrio vulnificus* and *Vibrio parahaemolyticus* in US retail shell oysters: A national survey from June 1998 to July 1999. *Journal of Food Protection,* 65(1): 79–87. See also correction, same authors, same journal, 65(3): 445.

Cook, D.W., Bowers, J.C. & DePaola, A. 2002. Density of total and pathogenic (*tdh+*) *Vibrio parahaemolyticus* in Atlantic and Gulf Coast molluscan shellfish at harvest. Journal of Food Protection, 65: 1873–1880.

Covert, D. & Woodburne, M. 1972. Relationships of temperature and sodium chloride concentration to the survival of *Vibrio parahaemolyticus* in broth and fish homogenate. *Applied Microbiology,* 23: 321–325.

Croci, L., Serratore, P., Cozzi, L., Stacchini, A., Milandri, S., Suffredine, E. & Toti, L. 2001. Detection of *Vibrionaceae* in mussels and in their seawater growing area. *Letters in Applied Microbiology,* 32: 57–61.

Croci, L., Suffredinin, E., Cozzi, L. & Toti, L. 2002. Effects of depuration of molluscs experimentally contaminated with *Escherichia coli, Vibrio cholerae* O1 and *Vibrio parahaemolyticus*. *Journal of Applied Microbiology,* 92: 460–465.

Dadisman, T.A. Jr., Nelson, R., Molenda, J.R. & Garber, H.J. 1972. *Vibrio parahaemolyticus* gastroenteritis in Maryland. I. Clinical and epidemiologic aspects. *American Journal of Epidemiology,* 96: 414–418.

Daniels, N.A., MacKinnon, L., Bishop, R., Altekruse, S., Ray, B., Hammond, R.M., Thompson, S., Wilson, S., Bean, N.H., Griffin, P.M. & Slutsker, L. 2000. *Vibrio parahaemolyticus* infections in the United States, 1973-1998. *Journal of Infectious Diseases,* 181: 1661–1666.

Davey, G., Prendergast, J. & M. Eyles. 1982. Detection of *Vibrio cholerae* in oysters, water and sediment from the Georges River. *Food Technology in Australia,* 34: 334–336.

Davis, J.W. & Sizemore, R.K. 1982. Incidence of *Vibrio* species associated with blue crabs (*Callinectes sapidus*) collected from Galveston Bay, Texas. *Applied and Environmental Microbiology,* 43: 1092–1097.

Deepanjali A., Sanath Kumar, H., Karunasagar, I. & Karunasagar, I. 2005. Seasonal variation in abundance of total and pathogenic *Vibrio parahaemolyticus* bacteria in oysters along the Southwest coast of India. *Applied and Environmental Microbiology,* 71: 3575–3580.

Degner, R.L. & C. Petrone. 1994. Consumer and restaurant manager reaction to depurated oysters and clams. Report to the Levy County Board of County Commissioners by the Florida Agricultural Market Research Center, University of Florida, USA.

Degner, R.L. 1998. Letter to S. Chirtel FDA/CFSAN on 5-Year Menu Census 1982–1987.

DePaola, A. 1999. Oyster cooling rates. Unpublished data.

DePaola, A. 2002. Oyster weights. Personal communication.

DePaola, A., Hopkins, L.H. & McPhearson, R.M. 1988. Evaluation of four methods for enumeration of *Vibrio parahaemolyticus. Applied and Environmental Microbiology,* 54: 617–618.

DePaola, A., Hopkins, L.H., Peeler, J.T., Wentz, B. & McPhearson, R.M. 1990. Incidence of *Vibrio parahaemolyticus* in U.S. coastal waters and oysters. *Applied and Environmental Microbiology,* 56: 2299–2302.

DePaola, A., Kaysner, C.A., Bowers, J. & Cook, D.W. 2000. Environmental investigation of Vibrio *parahaemolyticus* in oysters following outbreaks in Washington, Texas and New York (1997–1998). *Applied and Environmental Microbiology,* 66: 4649–4654.

DePaola, A., Kaysner, C.A., Nordstrom, J.L., Blackstone, G.M., Vickery, M. & Bowers, J.C. 2002. Harvest practices and ecological factors affecting the risk of *Vibrio parahaemolyticus* in Pacific Northwest oysters. Draft report.

DePaola, A., Nordstrom, J.L., Bowers, J.C., Wells, J.G. & Cook, D.W. 2003. Seasonal abundance of total and pathogenic *Vibrio parahaemolyticus* in Alabama oysters. *Applied and Environmental Microbiology,* 69: 1521–1526.

DePaola, A., Nordstrom, J.L., Blackstone, G.M., Bowers, J.C., Phillips, A.M.B., Grimes, D.J., Watt, S. & Parveen, S. 2007. Enumeration of total and pathogenic *Vibrio parahaemolyticus* using real-time PCR in a most probable number format. 6th International Conference on Molluscan Shellfish Safety. Blenheim, New Zealand, 18–23 March 2007.

Desenclos, J.A., Klontz, K.C., Wolfe, L.E. & Hoechen, S. 1991. The risk of *Vibrio* illness in the Florida raw oyster eating population, 1981–1988. *American Journal of Epidemiology,* 134: 290–297.

Desmarchelier, P. 1978. *Vibrio parahaemolyticus* and other vibrios. *Food Technology in Australia,* 30: 339–345.

Desmarchelier, P. 2003. Pathogenic Vibrios. *In:* A.D. Hocking (editor). *Foodborne Micro-organisms of Public Health Significance.* Australian Institute of Food Science and Technology Inc., North Sydney, Australia.

Einarsson, H. 1994. Evaluation of a predictive model for the shelf-life of cod (*Gadus morhua*) fillets stored in two different atmospheres at varying temperatures. *International Journal of Food Microbiology,* 24: 93–102.

Entis, P. & Boleszczuk, P. 1983. Overnight enumeration of *Vibrio parahaemolyticus* in seafood by hydrophobic grid membrane filtration. *Journal of Food Protection,* 46: 783–786.

European Commission. 2001. Opinion of the Scientific Committee on Veterinary Measures relating to Public Health on *Vibrio vulnificus* and *Vibrio parahaemolyticus* in raw and undercooked seafood. Available at http://ec.europa.eu/food/fs/sc/scv/out45_en.pdf

Eyles, M.J. & Davey, G.R. 1984. Microbiology of commercial depuration of the Sydney rock oyster, *Crassostrea commercialis. Journal of Food Protection,* 47: 703–706.

Eyles, M., Davey, G. & Arnold, G. 1985a. Behaviour and incidence of *Vibrio parahaemolyticus* in Sydney rock oysters (*Crassostrea commercialis*). *International Journal of Food Microbiology,* 1: 327–334.

Eyles, M., Davey, G., Arnold, G. & Wane, H. 1985b. Evaluation of methods for enumeration and identification of *V. parahaemolyticus* in oysters. *Food Technology in Australia,* 37(7): 302–304.

FAO/WHO. 2001. Report of the Joint FAO/WHO Expert Consultation on Risk Assessment of Microbiological Hazards in Foods: Hazard identification, exposure assessment and hazard characterization of *Campylobacter* spp. in broiler chickens and *Vibrio* spp. in seafoods. WHO Headquarters, Geneva, Switzerland, 23-27 July 2001. Available at: http://www.who.int/foodsafety/publications/micro/en/july2001_en.pdf

FAO/WHO. 2006. Report of the Joint FAO/WHO Expert Consultation on the use of microbiological risk assessment outputs to develop practical risk management strategies: Metrics to improve food safety. Kiel, Germany, 3–7 April 2006. Available at: ftp://ftp.fao.org/ag/agn/food/kiel.pdf

FDA [US Food and Drug Administration]. 2005. Quantitative Risk Assessment on the Public Health Impact of Pathogenic *Vibrio parahaemolyticus* in Raw Oysters. U.S. Department of Health and Human Services, U.S. Food and Drug Administration. See: http://www.fda.gov/Food/ScienceResearch/ResearchAreas/RiskAssessmentSafetyAssessment/ucm185746.htm

FDA/ISSC. 2000. *Vibrio vulnificus* and *Vibrio parahaemolyticus* in retail shell oysters – a national survey, June 1998-July 1999. Published as Cook et al., 2002, *q.v.*

FDA/ISSC. 2001. Density of *Vibrio parahaemolyticus* in shellfish at harvest, March 1999–September 2000.

Finkel, A. 1990. A simple formula for calculating the "Mass Density" of a lognormally distributed characteristic: applications to risk analysis. *Risk Analysis* 10(2):291–301.

Fisher, W.S. & DiNuzzo, A.R. 1991. Agglutination of bacteria and erythrocytes by serum from six species of marine molluscs. *Journal of Invertebrate Pathology*, 57: 380–394.

Fletcher, G. 1985. The potential food poisoning hazard of *Vibrio parahaemolyticus* in New Zealand oysters. *New Zealand Journal of Marine and Freshwater Research*, 19: 495–505.

Frischer, M.E., Thurmond, J.M. & Paul, J.H. 1990. Natural plasmid transformation in a high-frequency-of-transformation marine *Vibrio* strain. *Applied and Environmental Microbiology*, 56: 3439–3444.

Fuenzalida, L., Hernandez, C., Toro, J., Rioseco, M.L., Romero, J. & Espejo, R.T. 2006. *Vibrio parahaemolyticus* in shellfish and clinical samples during two large epidemics of diarrhoea in southern Chile. *Environmental Microbiology*, 8: 675–683.

Fujino, T., Okuno, Y., Nakada, D., Aoyoma, A., Fukai, K., Mukai, T. & Ueho, T. 1953. On the bacteriological examination of shirasu food poisoning. *Medical Journal of Osaka University*, 4: 299–304.

Garay, E., Arnau, A. & Amaro, C. 1985. Incidence of *Vibrio cholerae* and related vibrios in a coastal lagoon and seawater influenced by lake discharges along an annual cycle. *Applied and Environmental Microbiology*, 50: 426–430.

Geneste, C., Dab, W., Cabanes, P.A., Vaillant, V., Quilici, M.L. & Fournier, J.M. 2000. Les vibrioses non-cholériques en France : cas identifíes de 1995 à 1998 par le Centre National de Référence. *BEH [Bulletin Epidémiologique Hebdomadaire]*, 9: 38–40.

Genthner, F.J., Volety, A.K., Oliver, L.M. & Fisher, W.S. 1999. Factors influencing *in vitro* killing of bacteria by haemocytes of the Eastern oyster (*Crassostrea virginica*). *Applied and Environmental Microbiology*, 65: 3015–3020.

Gjerde, E.P. & Bøe, B. 1981. Isolation and characterisation of *Vibrio alginolyticus* and *Vibrio parahaemolyticus* from the Norwegian coastal environment. *Acta Veterinaria Scandinavica*, 22: 331–343.

Gonzalez-Escalona, N., Cachicas, V., Acevedo, C., Rioseco, M.L., Vergara, J.A., Cabello, F., Romero, J. and Espejo, R.T. 2005. *Vibrio parahaemolyticus* diarrhea, Chile, 1998 and 2004. *Emerging Infectious Diseases,* 11: 129–131.

Gonzalez-Escalona, N., Blackstone, G.M. & DePaola, A. 2006. Characterisation of a *Vibrio alginolyticus* strain isolated from Alaskan oysters carrying a gene similar to the thermostable direct hemolysin-related hemolysin gene (*trh*) of *Vibrio parahaemolyticus. Applied and Environmental Microbiology,* 72: 7925–7929.

Gooch, J., DePaola, A., Bowers, J. & Marshall, D. 2002. Growth and survival of *Vibrio parahaemolyticus* in postharvest American oysters. *Journal of Food Protection,* 65: 970–974.

Gregoire, A., Gras, R. & Khalanski, M. 1993. Surveillance hydrologique des centrals nucleaires implantees sur le littoral francais. *Hydroecologie Appliquee,* 5: 29–43.

Hally, R.J., Rubin, R.A., Fraimow, H.S. & Hoffman-Terry, M.L. 1995. Fatal *Vibrio parahaemolyticus* septicaemia in a patient with cirrhosis: a case report and review of the literature. *Digestive Diseases and Sciences,* 40: 1257–1260.

Hara-Kudo, Y., Sugiyama, K., Nishibuchi, M., Chowdhury, A., Yatsuyanagi, J., Ohtomo, Y., Saito, A., Nagano, H., Nishina, T., Nakagawa, H., Konuma, H., Miyahara, M. & Kumagai, S. 2003. Prevalence of pandemic thermostable direct haemolysin-producing *Vibrio parahaemolyticus* O3:K6 in seafood and the coastal environment in Japan. *Applied and Environmental Microbiology,* 69: 3883–3891.

Hariharan, H., Giles, J.S., Heaney, S.B., Arsenault, G., McNair, N. & Rainnie, J. 1995. Bacteriological studies on mussels and oysters from six river systems in Prince Edward Island. *Canadian Journal of Shellfish Research,* 14: 527–532.

HHS [US Department of Health and Human Services]. 1995. National Shellfish Sanitation Program Manual of Operations: Part 1. U.S. Department of Health and Human Services, Washington DC, USA.

Hlady, W.G. 1997. *Vibrio* infections associated with raw oyster consumption in Florida, 1981–1994. *Journal of Food Protection,* 60: 353–357.

Hlady, W.G. & Klontz, K.C. 1996. The epidemiology of *Vibrio* infections in Florida, 1981–1993. *Journal of Infectious Diseases,* 173: 1176–1183.

Honda, T., Ni, Y.X. & Miwatani, T. 1988. Purification and characterization of a haemolysin produced by a clinical isolate of Kanagawa phenomenon-negative *Vibrio parahaemolyticus* and related to the thermostable direct haemolysin. *Infection and Immunity,* 56(4): 961–965.

Hornstrup, M.K. & Gahrn-Hansen, B. 1993. Extraintestinal infections caused by *Vibrio parahaemolyticus* and *Vibrio alginolyticus* in a Danish county, 1987-1992. *Scandinavian Journal of Infectious Diseases,* 25(6): 735–740.

Hurley, C., Quirke, A., Reen, F. & Boyd E. 2006. Four genomic islands that mark post-1995 pandemic *Vibrio parahaemolyticus* isolates. *BMC Genomics,* 7: Article No. 104.

Ichige, A., Matsutani, S., Oishi, K. & Mizushima, S. 1989. Establishment of gene transfer systems for and construction of the genetic map of a marine *Vibrio* strain. *Journal of Bacteriology,* 171: 1825–1834.

Iida, T., Park, K.-S., Suthienkul, O., Kozawa, J., Yamaichi, Y., Yamamoto, K. & Honda, T. 1998. Close proximity of the *tdh, trh* and *ure* genes on the chromosome of *Vibrio parahaemolyticus. Microbiology,* 144: 2517–2523.

ICMSF [International Commission on Microbiological Specifications for Foods]. 1996. *Microorganisms in Foods 5. Characteristics of Microbial Pathogens.* Blackie Academic & Professional, London, UK.

ISSC & FDA (editors). 1997. National Shellfish Sanitation Program Guide for the Control of Molluscan Shellfish, Washington, D.C., USA.

Jackson, H. 1974. Temperature relationships of *Vibrio parahaemolyticus*. pp. 139–145, *In:* T. Fujino, G. Sakaguchi, R. Sakazaki and Y. Takeda (editors). International Symposium on *Vibrio parahaemolyticus*. Saikon Publishing Company, Tokyo, Japan.

Joseph, S.W., Colwell, R.R. & Kaper, J.B. 1982. *Vibrio parahaemolyticus* and related halophilic vibrios. *CRC Critical Reviews in Microbiology*, 10: 77–124.

Kaneko, T. & Colwell, R.R. 1975. Absorption of *Vibrio parahaemolyticus* onto chitin and copepods. *Applied and Environmental Microbiology*, 29: 269–274.

Kaneko, T. & Colwell, R.R. 1977. The annual cycle of *Vibrio parahaemolyticus* in Chesapeake Bay. *Microbial Ecology*, 4: 135–155.

Kaufman, G.E., Bej, A.K., Bowers, J. & DePaola, A. 2003. Oyster-to-oyster variability in the levels of *Vibrio parahaemolyticus*. *Journal of Food Protection*, 66: 125–129.

Kaysner, C.A. & Weagant, S.D. 1982. The incidence and seasonal distribution of *Yersinia enterocolitica* and *Vibrio parahaemolyticus* in a Puget Sound commercial oyster bed. *FDA-SARAP Report*, No. 104-79: 171–220.

Kaysner, C.A., Tamplin, M.L., Wekell, M.M., Stott, R.F. & Colburn, K.G. 1989. Survival of *Vibrio vulnificus* in shellstock and shucked oysters (*Crassostrea gigas* and *Crassostrea virginica*) and effects of isolation medium on recovery. *Applied and Environmental Microbiology*, 55: 3072–3079.

Kaysner, C.A., Abeyta, C., Stott, R.F., Krane, M.H. & Wekell, M.M. 1990a. Enumeration of *Vibrio* species including *V. cholerae* from samples of an oyster growing area, Grays Harbor, Washington. *Journal of Food Protection*, 53: 300–302.

Kaysner, C.A., Abeyta, C., Stott, R.F., Lilja, J.L. & Wekell, M.M. 1990b. Incidence of urea-hydrolyzing *Vibrio parahaemolyticus* in Willapa Bay, Washington. *Applied and Environmental Microbiology*, 56: 904–907.

Kelly, M.T. 1999. Personal communication regarding pathogenic *Vibrio parahaemolyticus*.

Kelly, M.T. & Stroh, E.M.D. 1988a. Occurrence of *Vibrionaceae* in natural and cultivated oyster populations in the Pacific Northwest. *Diagnostic Microbiology and Infectious Disease*, 9(1): 1–5

Kelly, M.T. & Stroh, E.M.D. 1988b. Temporal relationship of *Vibrio parahaemolyticus* in patients and the environment. *Journal of Clinical Microbiology*, 26: 1754–1756.

Kiiyukia, C., Venkateswaran, K., Navarro, I.M., Nakano, H., Kawakami, H. & Hashimoto, H. 1989. Seasonal distribution of *Vibrio parahaemolyticus* serotypes along the oyster beds in Hiroshima coast. *Journal of the Faculty of Applied Biological Sciences, Hiroshima University*, 28: 49–61.

Kishishita, M., Matsuoka, N., Kumagai, K., Yamasaki, S., Takeda, Y. & Nishibuchi, M. 1992. Sequence variation in the thermostable direct haemolysin-related haemolysin (*trh*) gene of *Vibrio parahaemolyticus*. *Applied and Environmental Microbiology*, 58: 2449–2457.

Klontz, K.C. 1990. Fatalities associated with *Vibrio parahaemolyticus* and *Vibrio cholerae* non-O1 infections in Florida (1981–1988). *Southern Medical Journal*, 83: 500–502.

Kothary, M.H., Tall, B.D., La Peyre, J.F., Shinaishin, S.J., Shah, D.B. & Faisal, M. 1997. *Perkinsus marinus* serine protease prolongs survival of *Vibrio vulnificus* in Eastern oyster haemocytes *in vitro*. Abstracts of the 97th General Meeting of the American Society for Microbiology. Abstract No. B-135:51.

Kothary, M.H., Burr, D.H., Tall, B.D., Hanes, D.E. & Miliotis, M.D. 2000. Thermostable direct haemolysin, thermostable-related haemolysin, and urease are not required for pathogenicity of *Vibrio parahaemolyticus* in animal models. Abstracts of the 100th General Meeting of the American Society for Microbiology. p. 68.

Kraa, E. 1995. Surveillance and epidemiology of foodborne illness in NSW, Australia. *Food Australia*, 47(9): 418–423.

La Peyre, J.F. & Volety, A.K. 1999. Modulation of Eastern oyster haemocyte activities by *Perkinsus marinus* extracellular proteins. Abstracts of the National Shellfisheries Association Annual Meeting. *Journal of Shellfish Research*, 18: 322.

Lawrence, D.N., Blake, P.A., Yashuk, J.C., Wells, J.G., Creech, W.B. & Hughes, J.H. 1979. *Vibrio parahaemolyticus* gastroenteritis outbreaks aboard two cruise ships. *American Journal of Epidemiology*, 109: 71–80.

Lake, R., Hudson, A. & Cressey, P. 2003. Risk profile: *Vibrio parahaemolyticus* in seafood. Institute of Environmental Science and Research Ltd, Christchurch, New Zealand.

Lemoine, T., Germanetto, P. & Giraud, P. 1999. Toxi-infection alimentaire collective à *Vibrio parahaemolyticus*. *BEH [Bulletin Epidémiologique Hebdomadaire]*, 10: 37–38.

Levine, W.C., Griffin, P.M. & Group, G.C.V.W. 1993. *Vibrio* infections on the Gulf Coast: results of first year of regional surveillance. *Journal of Infectious Diseases*, 167: 479–483.

Levy, A.S. & Fein, S.B. 1999. Trends in food safety cognitions and practices, 1993–1998. American Dietetic Association Annual Meeting, Atlanta, Georgia, USA.

Lewis, T., Brown, M., Abell, G., McMeekin, T. & Sumner, J. 2002. Pathogenic V*ibrio parahaemolyticus* in Australian oysters. Fisheries Research and Development Corporation Report, No. 2002/49.

Lindqvist, R., Andersson, Y., de Jong, B. & Norberg, P. 2000. A summary of foodborne disease incidents in Sweden, 1992 to 1997. *Journal of Food Protection*, 63: 1315–1320.

Liston, J. 1974. Influence of U.S. seafood handling procedures on V*ibrio parahaemolyticus*. pp. 123–128, *in:* T. Fujino, G. Sakaguchi, R. Sakazaki and Y. Takeda (editors). International Symposium on *Vibrio parahaemolyticus*. Saikon Publishing Company, Tokyo, Japan.

Long, J.S. & Freese, J. 2000. Scalar Measures of Fit for Regression Models. See: http://fmwww.bc.edu/repec/bocode/f/fitstat.pdf

McCarthy, S.A. & Khambaty, F.M. 1994. International dissemination of epidemic *Vibrio cholerae* by cargo ship ballast and other nonpotable waters. *Applied and Environmental Microbiology*, 60: 2597–2601.

McLaughlin, J.B., DePaola, A., Bopp, C.A., Martinek, K.A., Napolilli, N.P., Allison, C.G., Murray, S.L., Thompson, E.C., Bird, M.M. & Middaugh, J.P. 2005. Outbreak of *Vibrio parahaemolyticus* gastroenteritis associated with Alaskan oysters. *New England Journal of Medicine*, 353: 1463–1470.

Madigan, T., Lee, K., Pointon, A. & Thomas, C. 2007. A supply chain assessment of marine vibrios in Pacific oysters in South Australia: prevalence, quantification and public health risk. FRDC [Fisheries Research and Development Corporation] Project SIDF 2005/401.

Martinez-Urtaza, J., Simental, L., Velasco, D., DePaola, A., Ishibashi, M., Nakaguchi, Y., Nishibuchi, M., Carrera-Flores, D., Rey-Alvarez, C. & Pousa, A. 2005. Pandemic *Vibrio parahaemolyticus* O3:K6, Europe. *Emerging Infectious Diseases*, 11: 1319–1320.

Masini, L., De Grandis, G., Principi, F., Mengarelli, C. & Ottaviani, D. 2007. Research and characterisation of pathogenic vibrios from bathing waters along the Conero Riviera (Central Italy). *Water Research*, 41(18): 4031–4040

Mead, P.S., Slutsker, L., Dietz, V., McCaig, L.F., Bresee, J.S., Shapiro, C., Griffin, P.M. & Tauxe, R.V. 1999. Food-related illness and death in the United States. *Emerging Infectious Diseases*, 5: 607–625.

Miles, D.W., Ross, T., Olley, J. & McMeekin, T.A. 1997. Development and evaluation of a predictive model for the effect of temperature and water activity on the growth rate of *Vibrio parahaemolyticus*. *International Journal of Food Microbiology*, 38: 133–142.

Miwa, N., Nishio, T., Kawamori, F., Masuda, T. & Akiyama, M. 2003. Evaluation of MPN method combined with PCR procedure for detection and enumeration of *Vibrio parahaemolyticus* in seafood. *Shokuhin Eiseigaku Zasshi*, 44: 289–293.

Miyamoto, Y., Kato, T., Obara, Y., Akiyama, S., Takizawa, K. & Yamai, S. 1969. *In vitro* haemolytic characteristic of *Vibrio parahaemolyticus*: its close correlation to human pathogenicity. *Journal of Bacteriology*, 100: 1147–1149.

Motes, M.L. & DePaola, A. 1996. Offshore suspension relaying to reduce levels of *Vibrio vulnificus* in oysters (*Crassostrea virginica*). *Applied and Environmental Microbiology*, 62: 3875–3877.

Murphree, R.L. & Tamplin, M.L. 1991. Uptake and retention of *Vibrio cholerae* O1 in the Eastern oyster, *Crassostrea virginica*. *Applied and Environmental Microbiology*, 61: 3656–3660.

NACMCF [National Advisory Committee on Microbiological Criteria for Foods]. 1992. Microbiological criteria for raw molluscan shellfish. *Journal of Food Protection*, 55: 463–480.

Nair, G.B., Abraham, M. & Natarajan, R. 1980. Distribution of *Vibrio parahaemolyticus* in finfish harvested from Porto Novo (S. India) environs: a seasonal study. *Canadian Journal of Microbiology*, 26: 1264–1269.

Nair, G.B., Ramamurthy, T., Bhattacharya, S.K., Datta, B., Takeda, Y. & Sack, D.A. 2007. Global dissemination of *Vibrio parahaemolyticus* O3:K6 and its serovariants. *Clinical Microbiological Reviews*, 20: 39–48.

Nakaguchi, Y. & Nishibuchi, M. 2005. The promotor region rather than its downstream inverted repeat sequence is responsible for low level transcription of the thermostable direct haemolysin-related haemolysin (*trh*) gene of *Vibrio parahaemolyticus*. *Journal of Bacteriology*, 187: 1849–1855.

Natarajan, R., Abraham, M. & Nair, G.B.1980. Distribution of *Vibrio parahaemolyticus* in Porto Novo environment. *Indian Journal of Medical Research*, 71: 679–687.

Ng, T.C., Chiang, P.C., Wu, T.L. & Leu, H.S 1999. *Vibrio parahaemolyticus* bacteremia: case report. *Changgeng Yi Xue Za Zhi*, 22: 508–514.

Nishibuchi, M. 1999. *Vibrio parahaemolyticus* virulence. Personal communication.

Nishibuchi, M. & Kaper, J.B. 1995. Thermostable direct haemolysin gene of *Vibrio parahaemolyticus*: a virulence gene acquired by this bacterium. *Infection and Immunology*, 63: 2093–2099.

NMFS [National Marine Fisheries Service]. Various dates 1990 to 1998. Fisheries of the United States - (1998) from: www.nmfs.noaa.gov. NOT CITED

Nolan, C.M., Ballard, J., Kaysner, C.A., Lilja, J.L., Williams, L.P. & Tenover, F.C. 1984. *Vibrio parahaemolyticus* gastroenteritis. An outbreak associated with raw oysters in the Pacific Northwest. *Diagnostic Microbiology and Infectious Disease*, 2: 119–128.

Nordstrom, J.L., Vickery, M.C.L., Blackstone, G.M., Murray, S.L. & DePaola, A. 2007. Development of a multiplex real-time PCR assay with an internal amplification control for the detection of total and pathogenic *Vibrio parahaemolyticus* in oysters. *Applied and Environmental Microbiology*, 73(18): 5840–5847.

Ogawa, H., Tokunou, H., Kishimoto, T., Fukuda, S., Umemura, K. & Takata, M. 1989. Ecology of *V. parahaemolyticus* in Hiroshima Bay. *Journal of the Veterinary Association of Hiroshima Prefecture*, 4: 47–57.

Okuda, J., Ishibashi, M., Abbot, S.L., Janda, J.M. & Nishibuchi, M. 1997. Analysis of the thermostable direct haemolysin (*tdh*) gene and the tdh-related haemolysin (*trh*) genes in urease-positive strains of *Vibrio parahaemolyticus* isolated from West Coast of United States. *Journal of Clinical Microbiology*, 35: 1965–1971.

Oliver, J.D. 1995. The viable but non-culturable state in the human pathogen *Vibrio vulnificus*. *FEMS Microbiology Letters*, 133: 203–208.

Ordás, M.C., Novoa, B. & Figueras, A. 1998. Effect of cultured *Perkinsus atlanticus* supernatants on the mussel (*Mytilus galloprovincialis*) and carpet-shell clam (*Ruditapes decussatus*) defense mechanisms, p. 244, *in:* Proceedings of the Third International Symposium on Aquatic Animal Health.

Pan, T.M., Chai, T.-J., Lee, C.L., Chien, S.W. & Horng, C.B. 1997. Foodborne disease outbreaks due to bacteria in Taiwan, 1986 to 1995. *Journal of Clinical Microbiology*, 35: 1260–1262.

Parveen, S., Tamplin, M.L., da Silva, L.V.A., White, C., Bowers, J.C., Rutto, G. & DePaola, A. 2007. Predictive models for the growth and survival of total *Vibrio parahaemolyticus* in Gulf Coast shellstock oysters. Poster presentation at 94th Annual Meeting of the International Association for Food Protection (IAFP). Lake Bueno Vista, Florida, USA, 8–11 July 2007. pp. 176-177 in Abstracts.

Richards, G.P. 1988. Microbial purification of shellfish: A review of depuration and relaying. *Journal of Food Protection*, 51: 218–251.

Ristori, C.A., Iaria, S.T., Gelli, D. & Rivera, I.N.G. 2007. Pathogenic bacteria associated with oysters (*Crassostrea brasiliana*) and estuarine water along South Coast of Brazil. *International Journal of Environmental Health Research*, 17: 259–269.

Sakazaki, R., Iwanami, S. & Tamura, K. 1963. Studies on enteropathogenic facultatively halophilic bacterium, *Vibrio parahaemolyticus*. II. Serological characteristics. *Japanese Journal of Medical Science and Biology*, 21: 313–324.

Sanyal, S.C., Sil, J. & Sakazaki, R. 1973. Laboratory infection by *Vibrio parahaemolyticus*. *Journal of Medical Microbiology*, 6: 121–122.

Sanyal, S.C. & Sen, P.C. 1974. Human volunteer study on the pathogenicity of *Vibrio parahaemolyticus*, pp. 227–230, *in:* T. Fujino, G. Sakaguchi, R. Sakazaki and Y. Takeda. (editors). International Symposium on *Vibrio parahaemolyticus*. Saikon Publishing Company, Tokyo, Japan.

Sarkar, B.L., Nair, G.B., Banerjee, A.K. & Pal, S.C. 1985. Seasonal distribution of *Vibrio parahaemolyticus* in freshwater environs and in association with freshwater fishes in Calcutta. *Applied and Environmental Microbiology*, 49: 132–136.

Shirai, H., Ito, H., Hirayama, T., Nakamoto, Y., Nakabayashi, N., Kumagai, K., Takeda, Y. & Nishibuchi, M. 1990. Molecular epidemiologic evidence for association of thermostable direct haemolysin (TDH) and TDH-related haemolysin of *Vibrio parahaemolyticus* with gastroenteritis. *Infection and Immunity*, 58(11): 3568–3573

Son, N.T. & Fleet, G.H. 1980. Behavior of pathogenic bacteria in the oyster, *Crassostrea commercialis*, during depuration, re-laying, and storage. *Applied and Environmental Microbiology*, 40: 994–1002.

Sujatha, C.H., Nair, S.M. & Chacko, J. 1996. Tributyltin oxide induced physiological and biochemical changes in a tropical estuarine clam. Bull. Environ. *Bulletin of Environmental Contamination and Toxicology*, 56: 303–310.

Takikawa, I. 1958. Studies on pathogenic halophilic bacteria. *Yokohama Medical Bulletin*, 9: 313–322.

Tall, B.D., La Peyre, J.F., Bier, J.W., Miliotis, M.D., Hanes, D.E., Kothary, M.H., Shah, D.B. & Faisal, M. 1999. *Perkinsus marinus* extracellular protease modulates survival of *Vibrio vulnificus* in Eastern oyster (*Crassostrea virginica*) haemocytes. *Applied and Environmental Microbiology,* 65: 4261–4263.

Tamura, N., Kobayashi, S., Hashimoto, H. & Hirose, S-I. 1993. Reactive arthritis induced by *Vibrio parahaemolyticus. Journal of Rheumatology,* 20: 1062–1063.

TAS [TAS International Diet Research System]. 1995. USDA Nutrient Database. Data from 1989/90; 1990/91; 1991/92.

Tepedino, A.A. 1982. *Vibrio parahaemolyticus* in Long Island oysters. *Journal of Food Protection,* 45: 150–151.

Thieren, N. 1999. Washington State vibriosis cases. Symposium on Molluscan Shellfish Sanitation, Tacoma, USA.

Thompson, C.A., Vanderzant, C. & Ray, S.M. 1976. Serological and hemolytic characteristics of *Vibrio parahaemolyticus* from marine sources. *Journal of Food Science,* 41: 204–205.

Timbo, B.B., Altekruse, S.F., Headrick, M. & Klontz, K.C. 1995. Raw shellfish consumption in California: the 1992 California Behavioral Risk Factor Survey. *American Journal of Preventive Medicine,* 11(4): 214–217.

USDA [United States Department of Agriculture]. 1989–1992. Continuing Survey of Food Intake by Individuals. See: http://www.ars.usda.gov/Services/docs.htm?docid=7797#reports

Venkateswaran, K., Kiiyukia, C., Nakanishi, K., Nakano, H., Matsuda, O. & Hashimoto, H. 1990. The role of sinking particles in the overwintering process of *Vibrio parahaemolyticus* in a marine environment. *FEMS Microbiology Ecology,* 73: 159–166.

Volety, A.K., Genthner, F.J., Fisher, W.S., McCarthy, S.A. & Wiles, K. 1999. Differential effects of oyster (*Crassostrea virginica*) defences on clinical and environmental isolates of *Vibrio parahaemolyticus*. Abstract of the National Shellfisheries Association Annual Meeting *Journal of Shellfish Research,* 18: 326.

Wang, H., Wong, M., O'Toole, D., Mak, M., Wu, R. & Kong, R. 2006. Identification of a DNA methyltransferase gene carried on a pathogenicity island-like element (VPAI) in *Vibrio parahaemolyticus* and its prevalence among clinical and environmental isolates. *Applied and Environmental Microbiology,* 72: 4455–4460.

Washington State Department of Health. 2000. *Vibrio* monitoring data. Pers comm. from Ned Therien via FDA.

Washington State Department of Health. 2001. *Vibrio* monitoring data. Pers comm. from Charles Kaysner via FDA.

Watkins, W.D. 2000. Pacific Northwest oyster harvest times. Personal communication.

Watkins, W.D., Thomas, C.D. & Cabelli, V.J. 1976. Membrane filter procedure for enumeration of *Vibrio parahaemolyticus. Applied and Environmental Microbiology,* 32: 679–684.

Weinstein, J.E. 1995. Seasonal responses of the mixed-function oxygenase system in the American oyster, *Crassostrea virginica* (Gmelin 1791), to urban-derived polycyclic aromatic hydrocarbons. *Comparative Biochemistry and Physiology Part C: Pharmacology, Toxicology and Endocrinology,* 112: 299–307.

Wendt, P.H., Van Dolah, R.F., Bobo, M.Y., Mathews, T.D. & Levisen, M.V. 1996. Wood preservative leachates from docks in an estuarine environment. *Archives of Environmental Contamination and Toxicology,* 31: 24–37.

Xu, H-S., Roberts, N., Singleton, F.L., Attwell, R.W., Grimes, D.J. & Colwell, R.R. 1982. Survival and viability of nonculturable *Escherichia coli* and *Vibrio cholerae* in the estuarine and marine environment. *Microbial Ecology*, 8: 313–323.

Zhu, B.C.R., Lo, J.-Y., Li, Y.-T., Li, S.-C., Jaynes, J.M., Gildemeister, O.S., Laine, R.A. & Ou, C.-Y. 1992. Thermostable, salt tolerant, wide pH range novel chitobiase from *Vibrio parahaemolyticus*: isolation, characterization, molecular cloning, and expression. *Journal of Biochemistry*, 112: 163–167.

Appendix I-1
Impact of non-compliance in establishing limits for the level of *V. parahaemolyticus* in oysters

Calculation of effect of compliance rates

Based on Monte Carlo simulation results of the effect of limits with 100% compliance, the effect of less than 100% compliance to any given limit can be estimated using the equations:

$$HL(\alpha) = \alpha \times HL(1)$$

$$RR(\alpha) = \alpha \times RR(1)$$

Here, $HL(\alpha)$ = expected % harvest rejected and $RR(\alpha)$ = expected % risk reduction at compliance rate α ($0< \alpha<1$). Here, $HL(1)$ and $RR(1)$ correspond to the case $\alpha=1$ (i.e. 100% compliance). Given these formulae and values for $HL(1)$ and $RR(1)$ (i.e. predictions at 100% compliance), additional simulations to obtain predictions at less than 100% compliance are unnecessary.

Using the above formulae, predictions of risk reduction and harvest rejected for raw consumption are presented in Table I–A1.1 for each of the three action limits at compliance levels of less than 100%.

Justification of these formulae used to determine the predictions at less than 100% compliance is as follows:

Assume R, a random variable for the risk (per serving) under a scenario of α percent compliance, and let P be a random variable for whether or not the producer of a serving is complying with the given standard. Then (using the law of total probability) one can write the probability distribution of R under the scenario of α percent compliance as the weighted combination of conditional distributions of R when compliant to the given standard *versus* not compliant:

$$\Pr(R = r) = \Pr(R = r \mid P = 1) * \Pr(P = 1) + \Pr(R = r \mid P = 0) * \Pr(P = 0)$$

We know $\Pr(P=1)$ and $\Pr(P=0)$ by definition are equal to α and $1-\alpha$, respectively. $\Pr(R=r \mid P=0)$ is the distribution of baseline risk (i.e. when no standard is in effect, or there is a standard but with which no one complies). $\Pr(R=r \mid P=1)$ is the distribution of risk under the hypothetical case of complete adherence to the standard by all producers.

So, with E[.] denoting the expectation (average) of a random variable, we have:

$$RR(\alpha) = \frac{E[R \mid P=0] - E[R]}{E[R \mid P=0]} = \frac{E[R \mid P=0] - \left(\alpha \times E[R \mid P=1] + (1-\alpha) \times E[R \mid P=0]\right)}{E[R \mid P=0]}$$

$$= \frac{\alpha \times E[R \mid P=0] - \alpha \times E[R \mid P=1]}{E[R \mid P=0]} = \alpha \times RR(0)$$

The logic is the same for % harvest lost (*HL*). The distribution of risk (and other variables) under a compliance scenario is a mixture distribution. Here, it is a mixture of 2 distributions (100% compliance and C% compliance or baseline). The mixing ratio is the % compliance.

One must take note, of course, that the scheme (i.e. set of assumptions) considered here and leading to these simple formulae are somewhat simplistic. It is implicitly assumed that the compliance rate is independent of the level of hazard in the product. Also, there is no presumed risk of rejecting good product.

The effect of compliance rates to control of Vp/g at "cool-down" at action limits of 2, 3 and 4 logs is presented in Table I-A.1.

Table I–A1.1 Effect of compliance rates to control Vp/g at "cool-down" at action limits of 2, 3 and 4 Logs.

Country	Action level ($\log_{10}$)	Compliance rate (%)	Product rejected (%)	Risk reduction (%)
Australia	2	100	67.7	99.1
	3	100	21.1	87.3
	4	100	2.4	52.0
New Zealand	2	100	52.6	96.0
	3	100	9.5	66.3
	4	100	0.5	19.8
Japan	2	100	16.0	99.2
	3	100	4.7	96.8
	4	100	1.2	89.5
Australia	2	90	60.9	89.2
	3	90	19.0	78.5
	4	90	2.2	46.8
New Zealand	2	90	47.3	86.4
	3	90	8.5	59.7
	4	90	0.4	17.8
Japan	2	90	14.4	89.3
	3	90	4.2	87.1
	4	90	1.1	80.5
Australia	2	50	33.9	49.5
	3	50	10.6	43.6
	4	50	1.2	26.0
New Zealand	2	50	26.3	48.0
	3	50	4.7	33.2
	4	50	0.2	9.9
Japan	2	50	8.0	49.6
	3	50	2.3	48.4
	4	50	0.6	44.7

Appendix I-2
Evaluating the effectiveness of *V. parahaemolyticus* control measures

Patterns of the effectiveness of standards to reduce risk in comparison to the associated cost (i.e. harvest rejection) vary between the three countries (Japan, Australia and New Zealand) that were considered in response to the CCFH risk management question. These differences are predominantly a consequence of differences in predicted variability of log (Vp/g) at the control point. As the variability (in log (Vp/g)) increases, the distribution of exposure and risk becomes more inequitably distributed across consumers and a smaller percentage of the product is associated with a larger percentage of the illness burden. The variability of log(Vp/g) at the "cool-down" control point was predicted to be much higher for Japan than either Australia or New Zealand as a consequence of greater variability in post-harvest air temperatures and subsequent post-harvest growth.

A graphical summary of the inequity in a distribution is provided by the Lorenz curve, which is a plot of cumulative mass density versus cumulative probability percentile. When the distribution of a characteristic such as exposure or risk is lognormal, the mathematical form of the mass density is greatly simplified (Finkel, 1990) and the Lorenz curve is relatively easy to generate. Figure I–A2.1 shows Lorenz plots of predicted distributions of risk per serving for Japan, Australia and New Zealand. Curves based on lognormal approximations of the risk distributions are shown as dashed lines and the Monte Carlo simulation output of the effect of the standards that were considered explicitly (100/g, 1000/g, 10 000/g) are shown as symbols. The interpretation of the plot is that distributions with curves shifted toward the lower right corner are more "inequitable" than distribution with curves shifted towards the upper left corner. The more inequitable a distribution is, the more effective a standard will be in reducing risk with minimal cost (harvest rejection).

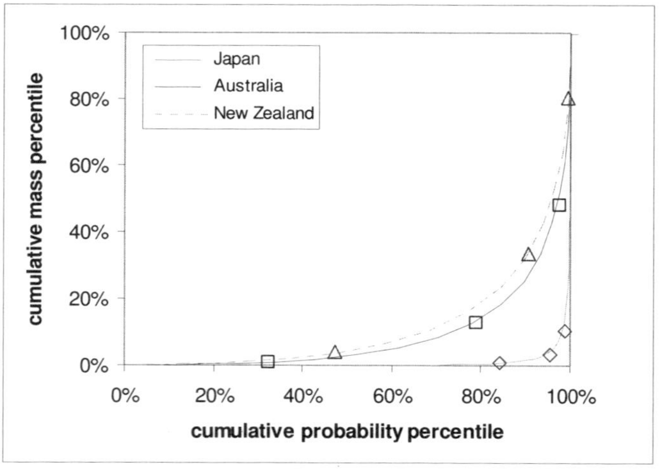

Figure A2.1 Median of Monte Carlo uncertainty distribution for (1-% harvest lost, 1-% risk reduction) plotted against Lorenz curve of a censored lognormal distribution with variance parameter equal to the median of the parameter's uncertainty distribution. Key: Triangle line = New Zealand, square line = Australia and diamond line = Japan.

Appendix I-3
Information on the model used for this assessment

The model used for the present risk assessment was that developed for the United States of America Quantitative Risk Assessment on the Public Health Impact of Pathogenic *Vibrio parahaemolyticus* in Raw Oysters (FDA, 2005). Full details of the model and how it can be operated are downloadable at http://www.cfsan.fda.gov/~dms/vpra-toc.html.

Both the parameters and the model may be changed or extended, giving it great potential for risk assessors to test a range of scenarios that might be important for them:

- Assess the effect of future global warming by increasing water and air temperatures.
- Test the effects of "perfect" oyster surveillance on the number of illnesses caused by oysters contaminated at specific levels.
- Localize harvest areas to specific *V. parahaemolyticus* densities and pathogenic *V. parahaemolyticus* densities.
- Invert the risk calculation: calculate the densities and prevalence of pathogenic in order to observe the present rate of illness.
- If salinity data are available, the model can be modified to include daily salinity variations in the calculation.
- If satellite data of water temperatures are available, the model could be run with real-time predictions of *V. parahaemolyticus* risk.
- Many different mitigations associated with harvesting practices can be explored by changing the model and seeing how sensitive the model is to the mitigation.

The model has utility to inform industry of the outcome of various "what-if" scenarios. An example of one such scenario—enhanced chilling of oysters after harvest from waters of different temperature (waters in different seasons)—is illustrated in Figure I–A3.1. This scenario was developed by the FDA-VPRA as part of the *V. parahaemolyticus* Control Plan, adopted by the General Assembly of the 2007 meeting of the Interstate Shellfish Sanitation Conference (ISSC).

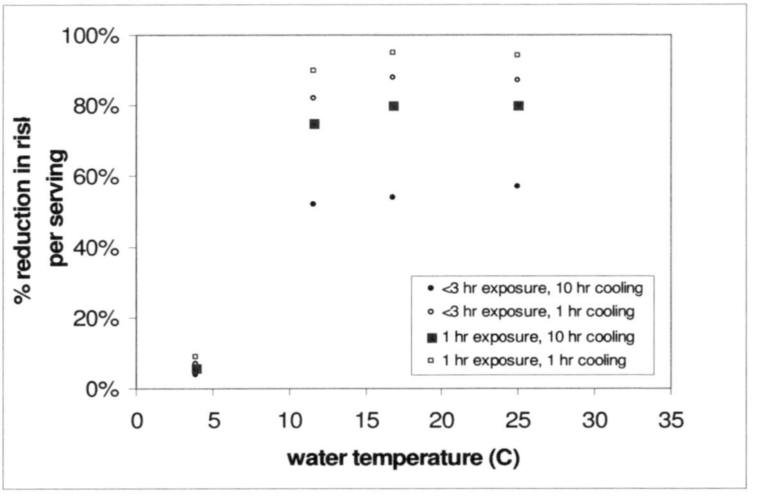

Figure I–A3.1. Effect of limiting ambient exposure prior to refrigeration and more rapid cooling on reduction of *V. parahaemolyticus* illnesses from consumption of oysters harvested at various water temperatures.

Figure I–A3.1 demonstrates the effectiveness of limiting ambient exposure prior to refrigeration and more rapid cooling on reduction of *V. parahaemolyticus* illnesses from consumption oysters harvested at various water temperatures. In general, the effectiveness of these measures increases with oysters harvested from warmer waters, as these are associated with warmer ambient air temperatures. With short exposure periods of 1 and 3 hours, rapid cooling contributed to the majority of risk reduction, which was >80% of the *status quo* (United States of America Mid-Atlantic) when oysters were chilled within 3 hours of harvest. Even placing oysters in refrigeration within 1 hour did not reduce illnesses by 80% at any of the temperatures under the 10-hour cooling scenario currently assumed in the risk assessment. Similar data for other regions is presented in Appendix 10 of the FDA-VPRA (FDA, 2005).

Under the *V. parahaemolyticus* Control Plan, states will have the option, when *V. parahaemolyticus* infections are likely to occur, of placing oysters under refrigeration within 5 hours of harvest or of using other methods to limit the its growth to a 0.75 log increase after harvest. This latter option may be attractive in situations where limiting harvest to 5 hours or on-board refrigeration are not practical.

The above example indicates the utility of the model as a practical tool to inform industry of processing outcomes without the need for conventional microbiology, which is expensive in time and resources.

Part II

Microbiological risk assessment of *Vibrio parahaemolyticus* in *Anadara granosa* (Bloody clam)

Comparison of two models:

"farm-to-fork" versus "fractional change"

II–1. Introduction

Vibrio parahaemolyticus is widely distributed in the marine environment of the temperate, subtropical and tropical zones of the world, where virulent strains can cause infection through consumption of contaminated seafood. In southern and eastern Asian countries, *V. parahaemolyticus* has been considered a major cause of foodborne gastroenteritis, perhaps reflecting the high level of seafood consumption in the culture of these countries (Fujino et al., 1974). In tropical and subtropical areas, as found in parts of Thailand, *V. parahaemolyticus* can be isolated from seawater and seafood throughout the year (Fujino et al., 1974). However, quantitative data on the occurrence of total as well as virulent strains of this organism are generally unavailable, except for raw oysters in the United States of America (Cook, Bowers and DePaola, 2002)

Few countries in the world have active foodborne disease surveillance systems, and most have no surveillance or national reporting system for vibrio infections other than cholera. However, health authorities have some information about the causes of diarrhoeal diseases through the investigation of outbreaks and, less often, through case-control studies of sporadic infections. Thailand lacks a formal reporting system that accurately reflects the incidence of *V. parahaemolyticus* infections; however, there have been several published reports of outbreak investigations (Anon., 2001c; Khuharat, 1998; Suthienkul, 2000; Tangkranakul et al., 2000) and the high percentage of sporadic illness (Echeverria et al., 1983). Based on this information, *V. parahaemolyticus* infections appear to be a significant public health concern in Thailand.

Researchers from Thailand and Japan developed two models of targeted risk assessment (a farm-to-fork model and a fractional change model) for this project, based on new data collected primarily from Hat Yai City in southern Thailand, where seafood consumption is popular.

II–2. Scope

The purpose of this targeted risk assessment was to estimate the risk of *V. parahaemolyticus* infection associated with consumption of seafood in a defined setting and during a limited period. This represents a case study in a developing country, where scientists were able to conduct a series of clinical and microbiological studies to generate locally relevant data and elaborate a risk assessment model for a non-oyster shellfish species.

II–3. Hazard identification

In eastern Asian countries and Japan, *V. parahaemolyticus* is a common cause of food-borne illness (Ministry of Health, Labour and Welfare, 1999–2001). The outbreaks caused by this organism occur frequently, but generally are small in size; outbreaks of less than ten cases are common.

In Hat Yai, a tropical town with a population of about 282 000, located between Songkla Lake (a brackish lake connected to the Gulf of Thailand) and the Andaman Sea, diarrhoeal illnesses due to *V. parahaemolyticus* were documented through a stool or rectal swab survey in two area hospitals. From 1999 to 2001, researchers were able to isolate *V. parahaemolyticus* from 504 samples.

A one-year culture survey was carried out in the two hospitals in 1999. The microbiological examination methods used are described in Appendix II–1. *V. parahaemolyticus* was isolated from 319 diarrhoeal specimens collected during the 1999 survey period: 276 (3.33%) of 8281 in Hat Yai Hospital and 43 (1.35%) of 3193 in Songklanagarind Hospital. Two strains isolated from two specimens became non-viable during storage, resulting in 317 viable *V. parahaemolyticus* isolates. Although the incidence of Shigella spp. was higher than that of *V. parahaemolyticus* in these specimens (Laohaprertthisan et al., 2003), the 319 sporadic cases in one year in two hospitals was considered very high for *V. parahaemolyticus* infection. The frequency of the isolation peaked in the summer season, from June to September. Based on data from the United States of America, approximately 1.2% of *V. parahaemolyticus* case patients are hospitalized (Mead et al., 1999); therefore the number of *V. parahaemolyticus* infections in the community would be expected to be much higher. Personal information was not collected on patients. This survey does not consider the time after onset of illness when cultures were taken, nor does it consider potential antibiotic use by individuals or in hospital. Both of these factors could result in further under-reporting of *V. parahaemolyticus* infections.

The 317 viable clinical strains isolated during the survey period were examined for the presence or absence of the toxin genes (*tdh* and *trh* genes), and their gene profiles are listed in Table II–1. The *tdh* gene was encountered in 91.5% of isolates, while isolates with the *trh* gene accounted for only 4.4% in all *V. parahaemolyticus* isolates examined. Thirteen isolates were positive for both *tdh* and *trh* (*tdh+ and trh+*). In one hospital, 64 clinical strains were isolated in 2000, and 123 in 2001. Pandemic strains (GS-PCR, *tdh+*, *trh-*) were prevalent in both 2000 (41 of 64 strains; 64%) and 2001 (83 of 123 strains; 68%). Among the serotypes of pandemic strains, the number of serotypes was limited to four in both years, with O3:K6 dominant (73% in 2000 and 76% in 2001), followed by O1:K25 (20% in 2000 and 13% in 2001) (Vuddhakul et al., 2006).

In order to identify pandemic strains among the 504 clincal strains of *V. parahaemolyticus* isolated during 1999–2001, which were initially characterized by the O3:K6 serovar, and have the *tdh* gene but not the *trh* gene (Okuda et al., 1997), the group-specific PCR (GS-PCR) assay, developed by Matsumoto et al. (2000), was applied to all *tdh+* strains. In total, 367 of the 504 strains (72.8%) were GS-PCR positive. Of these, 367 strains possessed the characteristics of pandemic strains (GS-PCR+ and *tdh*+), 285 strains were serovar O3:K6 and 41 strains belonged to O1:K25. Therefore, the pandemic strain O3:K6 serotype was dominant among all clinical isolates, followed by the pandemic O1:K25 serotype. The GS-PCR assay results for the 317 clinical strains isolated in 1999 are shown in Table II–1. Again, the predominant isolate is the pandemic O3:K6 serotype.

Table II–1. Characteristics of *Vibrio parahaemolyticus* strains isolated from diarrhoea patients in Hat Yai City, Thailand (1999) based on data was obtained for this work.

No. of strains	GS-PCR result	O:K Serovar*	Genes present *tdh*	Genes present *trh*	No. of strains	GS-PCR result	O:K Serovar*	Genes present *tdh*	Genes present *trh*
192	+	3:6	+	-	1	-	4:55	+	-
22	+	1:25	+	-	1	-	4:63	+	-
21	+	4:68	+	-	1	-	5:47	+	-
4	+	1:UT	+	-	1	-	6:18	+	-
1	+	1:41	+	-	1	-	8:22	+	-
1	+	4:12	+	-	1	-	12:UT	+	-
1	+	3:(6)**	+	-	1	-	1:UT	-	+
2	-	1:9	+	+	1	-	8:UT	-	+
2	-	1:UT	+	+	1	-	12:UT	-	+
1	-	1:41	+	+	1	-	13:UT	-	+
1	-	5:15	+	+	4	-	12:UT	-	-
1	-	5:UT	+	+	5	-	5:UT	-	-
1	-	12:UT	+	+	2	-	3:UT	-	-
6	-	4:8	+	-	1	-	1:UT	-	-
5	-	4:9	+	-	1	-	4:4	-	-
3	-	2:3	+	-	1	-	4:34	-	-
6	-	4:68	+	-	1	-	8:UT	-	-
2	-	1:56	+	-	1	-	10:24	-	-
2	-	4:UT	+	-	1	-	10:UT	-	-
2	-	8:22	+	-	2	-	11:19	-	-
2	-	8:41	+	-	1	-	11:UT	-	-
2	-	10:UT	+	-	1	-	12:53	-	-
1	-	3:7	+	-	1	-	UT:UT	-	-
1	-	4:4	+	-	1	-	R:UT	-	-

NOTES: * UT = untypeable; R = rough. ** Weak agglutination with anti-K6 serum

II–4. Available data

II–4.1 Food implicated

As far as it could be ascertained, there were no epidemiological or microbiological investigations of suspected food vehicles for the *V. parahaemolyticus* case patients in hospital, and no case-control study was undertaken. It was therefore not possible to demonstrate that any particular seafood was the source of *V. parahaemolyticus* infection in Hat Yai City. The prevalence of pathogenic strains in seafood or environmental samples was known to be relatively low (Wagatsuma, 1974; Matte et al., 1994; Wong, Ting and Shieh, 1992; Wong et al., 1999; Fang et al., 1987; Chowdhury et al., 2001). The fact that the proportion of pathogenic *V. parahaemolyticus* strains in those isolated from food and environmental samples, as expressed by the Kanagawa phenomenon (beta-type hemolysis on Wagatsuma agar), is reported as less than 2% (Sakazaki et al., 1968; Miyamoto et al., 1969), making it difficult to obtain important information for exposure assessment.

To increase the sensitivity of isolation procedures for obtaining exposure data, researchers applied a novel immunomagnetic separation method (Vuddhakul et al., 2000) specific for the O3:K6, O1:K25, O4:K68, O1:KUT (untypeable), O1:K41, and O4:K12 serotype strains common to the *V. parahaemolyticus* isolated from patients. The microbiological examination methods used are described in Appendix II-1. In brief, the method consisted of an enrichment culture followed by separation using immunomagnetic beads coated with antiserum specific to the above K antigens. From 1998 to 2000, 295 seafood samples were analysed: shrimp (50), crab (9), fish (100) and shellfish (136 bivalves). Only bivalves were found positive for virulent strains of *V. parahaemolyticus*: one in 1998, one in 1999 and two in 2000 (2.9%). Studies in 2001 therefore focused only on bivalves, and 9 virulent strains were isolated from 132 samples (6.8%). Serotype O3:K6 followed by O1:K25 were dominant among the thirteen virulent strains isolated (Table II–2). The O3:K6 and O1:K25 strains were GS-PCR+, *tdh+*, and *trh-*. It is possible that additional virulent serotypes may be present, but without the selective sensitivity of the immunomagnetic assay they would not be as readily isolated.

The common perception of the population of Hat Yai City is that consumption of Bloody clams is a major cause of diarrhoeal illness, including *V. parahaemolyticus* illnesses.

A comparison of the strains isolated from clinical and seafood samples presented in the hazard identification section indicated that bivalves, i.e. Bloody clam (*Anadara granosa*), Green mussel (*Perna viridis)* and Hard clam (*Meretrix lusoria*), from which virulent strains were isolated, were a potential source of *V. parahaemolyticus* infection in the Hat Yai area in 1998–2001 (Table II–2). As *Anadara granosa,* known locally as Bloody clam, is considered to be the most popular bivalve consumed in southern Thailand, it was selected as the first commodity in a risk assessment of *V. parahaemolyticus* in bivalves in Thailand.

Table II–2. Virulent strains of *V. parahaemolyticus* isolated from shellfish in Hat Yai City, Thailand (1998–2001). Data was obtained for this present report.

Strain no.	Month/year of isolation	Source*	Presence of gene		Serotype** result	GS-PCR
			tdh	*trh*		
PSU46	Dec/98	A	+	-	O3:K6	+
PSU47	Jan/99	A	-	+	O11:K3	-
PSU166	Aug/00	C	+	-	O3:K6	+
PSU228	Oct/00	B	+	-	O3:K6	+
PSU358	Apr/01	C	+	-	O3:K6	+
PSU359	Apr/01	B	+	-	O3:K6	+
PSU360	Apr/01	B	+	-	O3:K6	+
PSU434	May/01	B	+	-	O3:KUT	+
PSU435	May/01	B	+	-	O3:K6	+
PSU474	Aug/01	B	+	-	O3:K6	+
PSU476	Aug/01	C	+	-	O1:K25	+
PSU478	Dec/01	C	+	-	O1:K25	+
PSU513	Dec/01	A	-	+	O1:KUT	-

NOTES: *A = Bloody clam (*Anadara granosa*); B = Hard clam (*Meretrix lusoria*); C = Green mussel (*Perna viridis*); ** UT = untypeable.

The characteristics of pandemic serotype strains isolated from bivalves were identical to those pandemic serotype strains isolated from clinical specimens (Table II–1). In addition, the DNA fingerprints as demonstrated by the pulsed-field gel electrophoresis profiles of NotI-digested DNA of the O3:K6 and O1:K25 strains isolated from environmental and clinical sources were indistinguishable (Vuddhakul et al., 2006), which indicated the association between molluscan shellfish (including Bloody clam) consumption and *V. parahaemolyticus* infections in humans. Due to the lack of epidemiological investigations of patients concerning the attribution of food or their behaviour in this risk assessment, the level of contribution of Bloody clam to *V. parahaemolyticus* illness could not be defined, Thus, in this risk assessment, the contribution of Bloody clam to *V. parahaemolyticus* illnesses is an assumption. However, there is some biological rationale to the assumption that bivalves are associated with *V. parahaemolyticus* infections in this area, namely:

- there is a correlation between the pathogens found in clinical specimens and in bivalves, as described above;
- bivalves are filter-feeders and are known to accumulate to high internal concentrations bacterial and viral agents from seawater and estuarine environments;
- bivalves are reported to be vehicles of transmission of *V. parahaemolyticus* infections (Daniels et al., 2000a, b);
- seafood is an extremely popular meal in Thailand, and people generally undercook bivalves so that the meat remains soft; and

- there is a common perception among the population that diarrhoeal illness is associated with the consumption of bivalves. When 80 people who eat Bloody clam were interviewed (see Section 4.6 below: Consumption of Bloody clam), 38 people answered that they had experienced diarrhoea after consumption of Bloody clam (Yamamoto et al., 2008).

II–4.2 Prevalence of V. parahaemolyticus

The prevalence of total *V. parahaemolyticus* and that of *tdh*+ and *trh*+ strains of *V. parahaemolyticus* in Bloody clam was examined at harvest, at retail and post-cooking (boiling) stages (Table II–3). Thirty-two sample sets were examined. The same lots of Bloody clam were examined for the harvest, retail, and post-cooking stages for the same sampling set. The presence or absence of *V. parahaemolyticus* (hereafter termed total *V. parahaemolyticus*), *tdh*-positive *V. parahaemolyticus* (hereafter termed *tdh*+ *V. parahaemolyticus*), and *trh*-positive *V. parahaemolyticus* (hereafter termed *trh*+ *V. parahaemolyticus*) contained in each tube of the final enrichment culture (6-hour salt polymyxin broth culture) was examined using two methods: direct PCR and colony-isolation PCR. The MPNs of total *V. parahaemolyticus*, *tdh*+ *V. parahaemolyticus*, and *trh*+ *V. parahaemolyticus* were determined using both methods. The prevalence of total and virulent *V. parahaemolyticus* is shown as the simple rate of positive samples among all the samples tested. The exposure assessment model was developed based primarily on the PCR data presented in Table II–3.

The results obtained by the two methods partially agreed for total *V. parahaemolyticus*, but not for *tdh*+ *V. parahaemolyticus* and *trh*+ *V. parahaemolyticus*. The *tdh*+ *V. parahaemolyticus* were detected in four sets of the samples by the direct PCR method and in two sets by the colony-isolation PCR method at the harvest stage, and only the results obtained for one set agreed. The disagreement between the two methods is probably in part due to the low prevalence of *tdh*+ *V. parahaemolyticus* and *trh*+ *V. parahaemolyticus*. Overall, the total number of *tdh*+ *V. parahaemolyticus*- or *trh*+ *V. parahaemolyticus*-positive sample sets detected by the two methods were equal (seven), thus suggesting the sensitivities of the two methods were not very different. When the target organism is present at a low level, the direct PCR method is expected to be more sensitive than the colony-isolation-based method (Dileep et al., 2003). Therefore, the less laborious direct PCR method was used for the rest of the study.

Table II–3. Prevalence of total and virulent *Vibrio parahaemolyticus* in 32 samples of Bloody clam at harvest, retail and post-cooking at a single harvest site in Pattani Province in Southern Thailand.

	Direct PCR			**colony isolation-PCR**		
	Total Vp	*tdh*+ Vp	*trh*+ Vp	Total Vp	*tdh*+ Vp	*trh*+ Vp
Harvest	32 (100%)	4 (13%)	2 (6%)	32 (100%)	2 (6%)	3 (9%)
Retail	32 (100%)	0 (0%)	1 (3%)	32 (100%)	0 (0%)	2 (6%)
Post-cooking	1 (3%)	0 (0%)	0 (0%)	2 (6%)	0 (0%)	0 (0%)

Table II–4. Most probable number (MPN) of total *Vibrio parahaemolyticus* (Vp) and virulent *V. parahaemolyticus* (*tdh*+ or *trh*+) per Bloody clam at harvest and retail stages by the direct PCR method. Harvest and retail samples correspond to the same batch.

Batch-sample no.	Harvest Stage			Retail Stage		
	Total Vp	*tdh*+	*trh*+	Total Vp	*tdh*+	*trh*+
1	8	0	0	15	0	0
2	26	0	0	263	0	0
3-1	84	0	0	161	0	0
3-2	385	0	0	33	0	0
3-3	161	0	0	161	0	0
4-1	161	0	0	1 610	0	0
4-2	161	0	0	735	0	0
4-3	5	0	0	840	0	0
5-1	53	0	0	8 400	0	0
5-2	326	1	0	5 250	0	0
5-3	385	0	0	38 500	0	0
6-1	3 255	0	0	38 500	0	0
6-2	16 100	0	0	5 250	0	0
6-3	8 400	0	0	38 500	0	0
7-1	81	0	0	2 625	0	0
7-2	53	0	0	525	0	0
7-3	33	0	0	161	0	0
8-1	5 250	0	0	15 050	0	0
8-2	8 400	0	0	8 400	0	0
8-3	735	0	0	5 250	0	0
9-1	5 250	0	1	5 250	0	0
9-2	2 625	0	0	8 400	0	0
9-3	1 505	0	0	16 100	0	0
10-1	5 250	0	0	16 100	0	0
10-2	16 100	0	1	1 505	0	1
10-3	8 400	0	0	3 255	0	0
11-1	4 200	0	0	2 625	0	0
11-2	1 505	1	0	5 250	0	0
11-3	805	1	0	2 625	0	0
12-1	2 625	0	0	16 100	0	0
12-2	1 365	1	0	735	0	0
12-3	8 400	0	0	735	0	0
Total	102 090	4	2	248 908	0	1

II–4.3 Concentration of total and virulent V. parahaemolyticus at harvest and retail stages

In a data collection process, one lot of clams was obtained soon after landing from fishing vessels at a harvest site. After initial sampling ("Harvest" stage), the rest of the clams were transported to the area where the local market is usually open and the laboratory is also located. The clams were then examined ("Retail" stage). Thereafter, the clams were maintained outside of the laboratory to simulate the actual transportation conditions, and then sampled again for analysis. The typical way of cooking clams in the home is by boiling for a short period (insufficient heating in some cases).

The most probable number (MPN) of total *V. parahaemolyticus* and virulent strains of *V. parahaemolyticus* per shellfish at harvest and retail stages as determined by the direct PCR method are summarized in Table II–4. The MPNs of *tdh+ V. parahaemolyticus* and *trh+ V. parahaemolyticus* were one per shellfish in all of sample sets where *tdh+ V. parahaemolyticus* or *trh+ V. parahaemolyticus* was positive.

II–4.4 Growth during transportation from harvest to retail, and from retail to cooking

The time from harvest to retail was approximately 5 hours, which resulted in the change in bacteria numbers noted above between harvest and retail. The number of total *V. parahaemolyticus* at harvest and retail stages are compared in Figure II–1, which provides strong evidence that growth occurs.

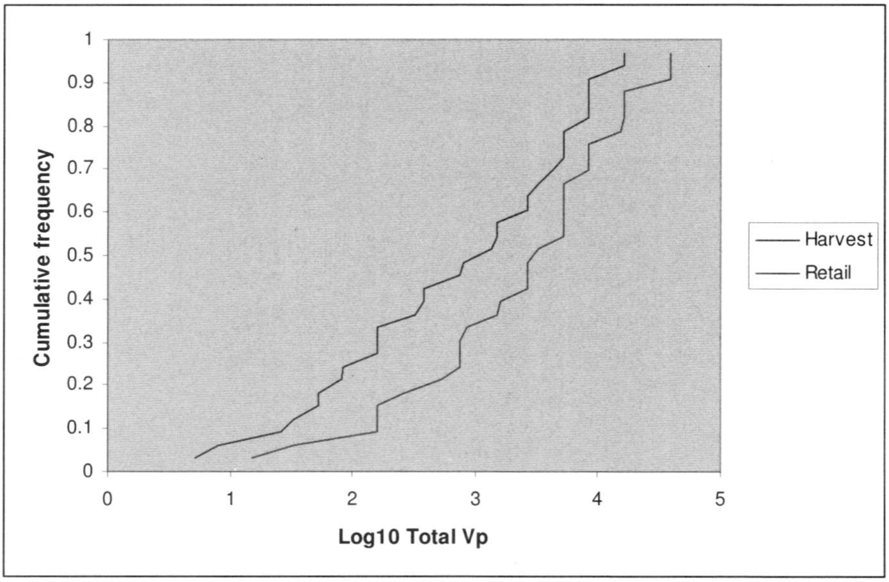

Figure II–1. Cumulative frequency plot of number of organisms per shellfish of total *Vibrio parahaemolyticus* at harvest and retail stages, showing growth.

The period between retail and pre-cooking was assumed to follow a PERT distribution (Vose, 2000) with a range of 0 to 25 hours, and a most likely value of 1 hour, based on observations of local individual practices and expert opinion. The PERT distribution is used exclusively for modelling expert estimates, where one is given the expert's minimum, most likely and maximum guesses. It is a direct alternative to a Triangle distribution. The PERT distribution is a smooth curve that puts less emphasis on extreme values, and places progressively more emphasis on values around the most likely value. It is often used in Monte Carlo simulations.

II–4.5 Cooking (boiling) process

People living in the study area were recruited to study actual cooking procedures. A total of 20 people boiled Bloody clam as they usually do at home and the temperature inside the clam meat was monitored throughout the boiling, using a wire probe connected to an automated recorder (Thermocouple, AP-810 Anritsu, Tokyo). As a result of measurements, 19 out of 20 people cooked enough to kill all bacteria, whereas one person boiled the clams with the internal temperature reaching a maximum of 50.5°C in a period of a few seconds. Clams were assumed to be thoroughly cooked if the internal temperature of the clam was kept at at least 64°C for more than 90 seconds The criterion for this assumption is based on a comprehensive reference book (ICMSF, 1996).

One obvious limitation of this study is that people, knowing that they were being monitored, may have taken more care to cook the clams properly.

II–4.6 Consumption of Bloody clam

The frequency of clam consumption, based on the interview survey, ranged from 2 to 20 occasions per year (Figure II–2).

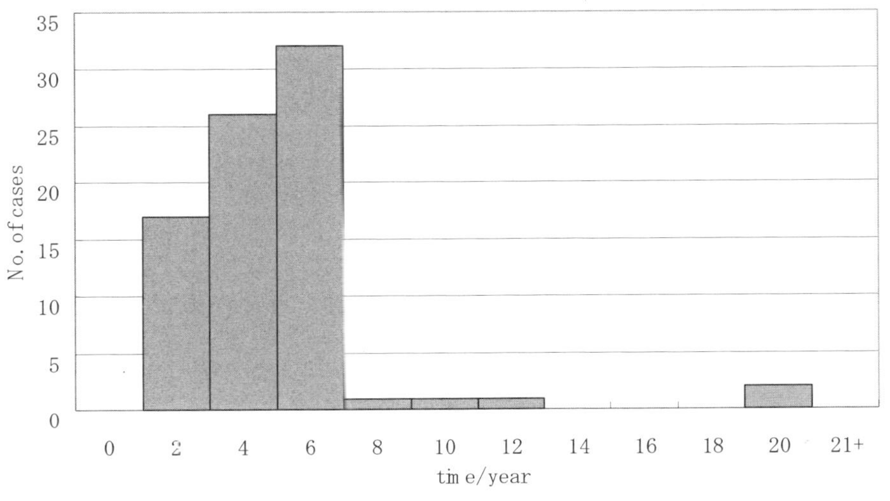

Figure II–2. Distribution of the frequency of consumption of Bloody clam per year.

The number of clams consumed per eating occasion ranged from 3 to 50 (Figure II–3). There seems to be a negative correlation between the number of clams in a meal and the number of clam meals per year consumed for the individuals surveyed (Figure II–4).

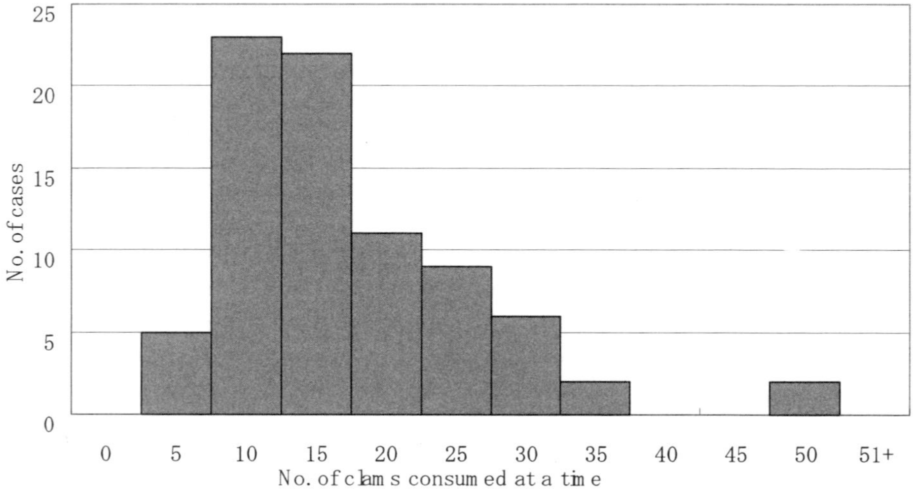

Figure II–3. Distribution of the number of Bloody clam consumed per eating occasion.

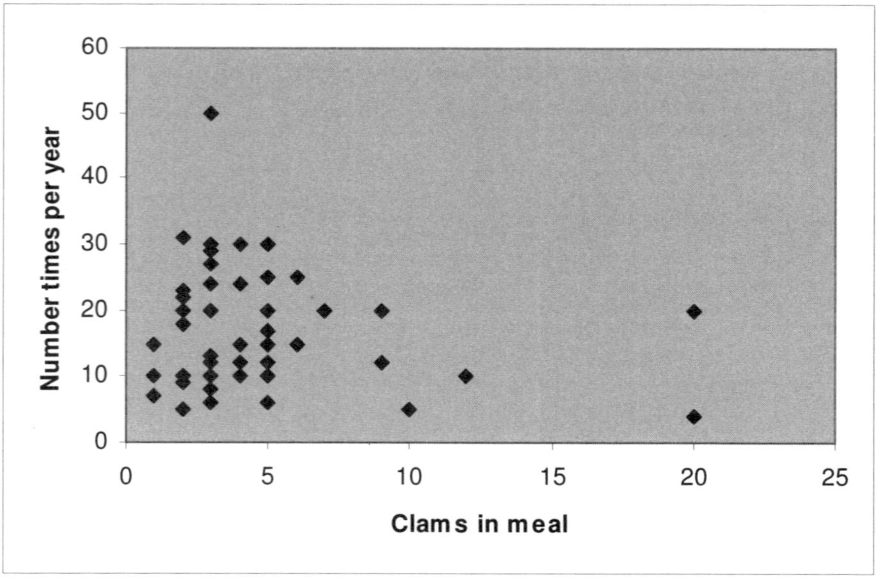

Figure II–4. Scatter plot of the number of clams in a meal versus the number of clam meals per year for the individuals surveyed. Although inconclusive, there may be a negative correlation.

II–4.7 Dose-response data

The USFDA (Anon., 2001b) fitted several dose-response models to the *V. parahaemolyticus* feeding trial data for the probability of illness, and found the Beta-Poisson model to be the best fit. The Beta-Poisson model is the only one that meets the mechanistic criteria (there is no threshold level, i.e. a single cell can cause illness) identified by FAO/WHO (2003). They estimated the uncertainty about the parameters α and β for Equation 4 in Appendix II–3 using the Bootstrap method (Haas, 1999). The result of their dose-response curve fitting is shown in Figure II–5. For the Beta-Poisson, α and β are the shape (steepness) and location parameters, respectively.

To include uncertainty related to parameters and maximum likelihood parameter estimated and their likelihood of Bootstrap outcome presented in *V. parahaemolyticus* risk assessment by USFDA (2005) were used.

The mean probability of a single cfu of *V. parahaemolyticus* causing illness (gastroenteritis or septicaemia) by this model is given by:

$$\bar{p} = \frac{\alpha}{\alpha + \beta}$$

Figure II–5 shows how the Beta-Poisson model FDA fitted has considerable uncertainty. However, what is apparent is that the product P (infection | dose) *P (illness | infection) is low for doses below 10^5 cfu and there is an approximate straight line relationship with ingested mean dose. Furthermore, the FDA report notes:

> "Consideration of the predicted density of pathogenic *V. parahaemolyticus*, the number of raw oyster servings for the Gulf Coast summer harvest and the likely number of illnesses occurring [Kennedy, as cited in Anon., 2001], strongly suggests that the predicted risks per serving based on dose-response curves ... are not plausible. Consequently, direct extrapolation of the dose-response under conditions of exposure in the feeding trials is not supported by the epidemiological data."

Later, the report also notes:

> "Given these estimates of annual illness rate it was determined that at least a 10-fold increase of the ID_{50} estimated with respect to the feeding trials was necessary to infer a dose-response consistent with the epidemiology. It is possible that the true ID_{50} for the general population is even greater than implied by this adjustment but this uncertainty was not evaluated in the present risk assessment."

The FDA risk assessment demonstrated that the feeding trial data were for some reason inconsistent with the United States of America experience and that the probability of infection should be some ten times smaller than the fitted model, or even less.

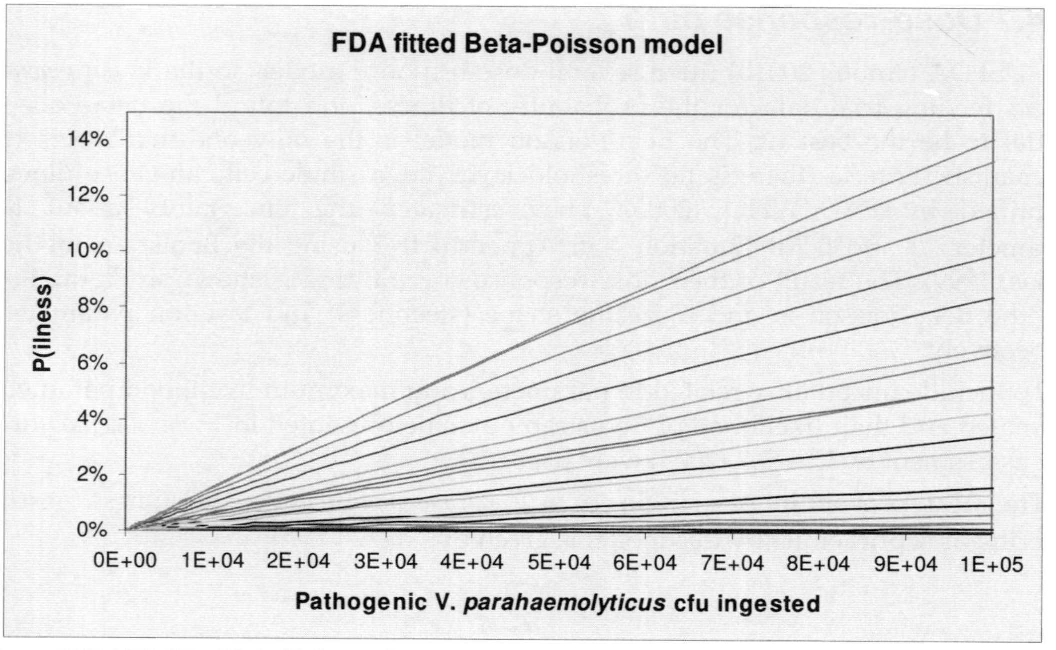

Figure II–5. FDA fitted Beta-Poisson dose-response model. Lines show fits to Bootstrap re-samples, which give an approximate uncertainty on the fitted model.

II–4.8 Summary of available data

1. There are no data to determine the annual incidence of *V. parahaemolyticus* illnesses in Thailand, though it is suspected to be quite high.
2. There is only weak casual evidence to link the consumption of Bloody clam with *V. parahaemolyticus* illnesses in humans in Thailand.
3. There are no case-control studies or other epidemiological evidence to determine the fraction of *V. parahaemolyticus* illnesses in Thailand that could be attributed to Bloody clam consumption.
4. Prevalence estimates are available for pathogenic *V. parahaemolyticus* in Bloody clam, but not for other possible exposure routes.
5. Data on the number of pathogenic *V. parahaemolyticus* in clams have been collected at harvest and retail, with the hours measured in between, which allows the estimation of growth rates.
6. Individual data have been collected for cooking habits and for consumption patterns of Bloody clam, but not for other food types.
7. Some weak subjective data are available for the length of time Bloody clam is cooked prior to consumption.
8. The dose-response data from the *V. parahaemolyticus* feeding trial do not appear to correspond well to human health effects in the real world, but we know that a linear relationship between dose and probability of illness, or expected number of illnesses, is reasonably robust.

II–5. Possible models that could be made with the available data and the questions they could answer

In performing a risk assessment that is intended to be useful to a decision-maker, efforts need to be made to ensure that any model adheres to the following criteria:

- it addresses as many questions as possible posed by the decision-maker, and in a timely fashion;
- it is supported by the best available data;
- it makes the fewest and most robust assumptions; and
- it is as efficient as possible.

One way to ensure that one does not embark on producing a risk assessment model that ultimately has no value to a decision-maker is to consider what data are available, or could be made available within a reasonable timeframe (where 'reasonable' is determined by the decision-maker). This section discusses two models that could be developed with the available data already presented here, and how those models might assist a decision-maker. Although there is considerable debate amongst the food safety community as to what constitutes the domain of the risk assessor and the risk manager, it is clearly quite critical that the risk assessor explains what can and cannot be done to help the risk manager with the information at hand.

II–5.1 Farm-to-fork style model

A model based on the parameters indicated in Figure II–6 was constructed from the available data. The weakest point in the model would probably be the dose-response relationship. It could be useful for decision-makers in offering some evidence as to whether Bloody clam is actually a major cause of *V. parahaemolyticus* illnesses in Thailand, and this concept is explored below.

Assume that all *V. parahaemolyticus* illnesses are caused by Bloody clam, and assume that all cases are sporadic, not outbreaks. Produce the risk assessment model to give an approximate prediction of the number of human cases that would result from the contaminated Bloody clam, taking into account cooking and consumption patterns. If this estimated incidence is approximately what the authorities would expect, the risk assessment provides some support for the contention that Bloody clam is the major cause of *V. parahaemolyticus* infections. If the calculated incidence is significantly higher than expected, the model is likely to be wrong. Finally, if the calculated incidence is significantly less than authorities believe occurs, this suggests (if the model is reasonably correct) that Bloody clam is not the major source of *V. parahaemolyticus* infections, or this model is not properly representing major routes of infections. For example, cross-contamination from Bloody clam might be an important route of infection.

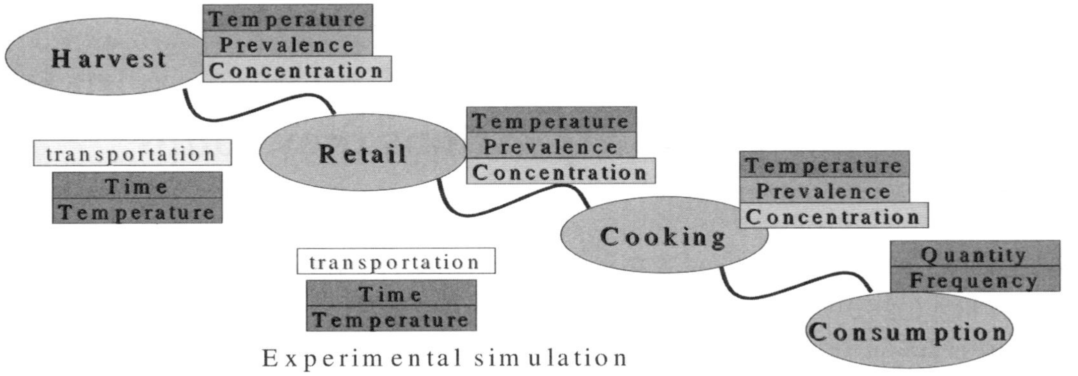

Figure II–6. Schematic representation of the model framework for a production-to-consumption risk assessment of *V. parahaemolyticus* in Bloody clam.

II–5.2 Estimation of fractional reduction in human cases of V. parahaemolyticus *due to changes in clam processing*

The approximate linear relationship between bacterial load and probability of illness means that the fractional change in microbial load or prevalence of *V. parahaemolyticus*, or both, that would occur in Bloody clam with a change in processing would result in the same fractional change of human illnesses, provided that the bacterial load in nearly all Bloody clam meals remained below 10^6 cfu. It is therefore quite possible to consider the fractional reduction in disease that would occur with a pathogen-reducing change in handling or storage of Bloody clam, provided that data are available to estimate that load reduction, and that this calculation can be made without the need to use a dose-response model, beyond making the linear assumption already discussed. Such an analysis is of quite limited value for the management of *V. parahaemolyticus* risk from Bloody clam, but does show the risk manager that if an epidemiological study were to be conducted to estimate the number of *V. parahaemolyticus* illnesses caused by Bloody clam, this risk assessment could provide information on the health impact of various risk mitigation strategies focused on reducing the amount of *V. parahaemolyticus* exposure.

II–6. Farm-to-fork style model

A model of the form of that in Figure II–6 was constructed with the available data, using Monte Carlo simulation in the @RISK/Excel modelling environment. The model logic proceeded is considered below.

1. The proportions of all *V. parahaemolyticus* cfu in Bloody clam that were virulent strains, either *tdh*+ or *trh*+ strains, (p_{tdh}+, p_{trh}+) were estimated. It was assumed that these proportions remained statistically constant throughout the harvest to consumption continuum, which effectively means assuming no difference between virulent and non-virulent strain survivability. It was also assumed that all Bloody clams contain some non-zero level of *V. parahaemolyticus*.
2. The distribution of total *V. parahaemolyticus* cfu in clams N_r at retail was estimated from sample data.
3. The ratio of the total *V. parahaemolyticus* concentration at retail to that at harvest was used, together with knowledge of the time between harvest and retail, to model *V. parahaemolyticus* growth rate *k* and the total number of *V. parahaemolyticus* bacteria at retail.
4. The concentration of total *V. parahaemolyticus* at retail was elevated by a growth factor based on the *subjectively* estimated time clams are stored post-retail before consumption, together with *k*, p_{tdh}+ and p_{trh}+ to estimate the number of pathogenic cfu in a random clam at the point of consumption.
5. The fraction of Bloody clam meals that are improperly cooked was estimated from a small survey. All *V. parahaemolyticus* cfu were assumed to survive improper cooking, and assumed to become inactive with proper cooking.
6. Survey data on the number of clams eaten in a meal was used together with steps 4 and 5 above to model the number of pathogenic *V. parahaemolyticus* consumed in a random meal.
7. Individual annual consumption was estimated from survey data. Possible correlation between the clams eaten in a meal and the number of clam meals per year visible in the survey data was preserved in the simulated scenarios.
8. The probability of illness from a random simulated meal was calculated using the Beta-Poisson dose-response model. The total pathogenic dose was estimated by adding the simulated number of *tdh*+ and *trh*+ cfu in the meal. This makes the assumption, in the absence of contradictory information, that the strains are equally virulent.
9. The probability of illness from a random meal size (in terms of clams consumed) was weighted by the probability of having such a sized meal. Integrating this probability over all simulated meals gives the expected number of times a person will become ill in a year with *V. parahaemolyticus* from Bloody clam.

These steps and their assumptions are described in more technical detail in the following sections.

II–6.1 Proportions of all V. parahaemolyticus *cfu in Bloody clam that were virulent strains* (tdh+ *or* trh+ *strains)*

The fractions of all *V. parahaemolyticus* cfu in Bloody clams that are *tdh+* or *trh+* strains (p_{tdh}+, p_{trh}+) were estimated in the same manner using Bayesian inference with a Uniform (0,1) uninformed prior, and a binomial likelihood function based on the assumptions that:

- any one random *V. parahaemolyticus* cfu has the same probability of being either *tdh+* or *trh+*;
- the direct PCR data from Table II–4 showing the presence or absence of *tdh+* and *trh+* strains has 100% sensitivity and specificity; and
- the probabilities that a cfu is *tdh+* or *trh+* are independent of each other.

The calculation resulted in the estimates shown in Figure II–7.

Assumptions

The binomial approximation and separately estimating the fractions p_{tdh}+ and p_{trh}+ are robust if the two fractions are small (as they are).

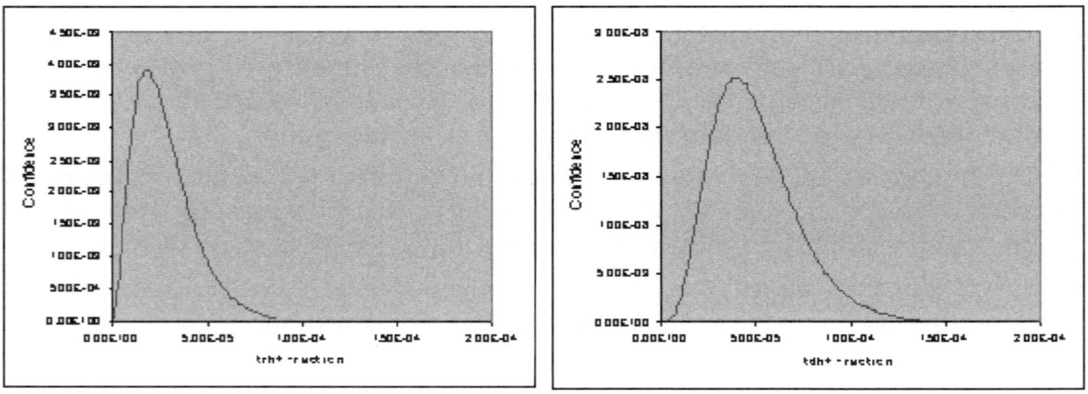

Figure II–7. Estimates of the fraction of *V. parahaemolyticus* cfu in Bloody clam that are *trh+* and *tdh+* strains.

II–6.2 Total V. parahaemolyticus *cfu in clams at retail (N_r)*

The $\log_{10}$ of the *V. parahaemolyticus* cfu counted at harvest in Table II–4 fit reasonable well to a Normal distribution with mean 3.345 and standard deviation 0.869, which is equivalent to a Lognormal distribution of cfus:

$$N_r = 10\text{^}Normal(3.345, 0.869)$$

Uncertainty about the mean and standard deviation were modelled using the Student-t and Chi-square distributions as appropriate for the Normal distribution assumption (Vose, 2000).

II–6.3 V. parahaemolyticus growth rate with temperature

The simple environment-independent exponential growth model was used, in the absence of a better *V. parahaemolyticus* growth model:

$$N_r = N_h e^{kt}$$

where N_h is the initial number of *V. parahaemolyticus* cfu at harvest, N_r is the number of cfu at retail, t is the time taken between harvest and retail, and k is the growth rate for a given temperature. In Log_{10} units this gives:

$$k = (Log_{10}(N_r) - Log_{10}(N_h)) * Log_e(10)/t$$

Using this formula with the data from Table II–4, and recognizing that the time from harvest to retail is approximately t = 5 hours, the growth rate *k* was assumed to be given by a Normal distribution with mean of 0.236 and standard deviation of 0.390:

$$k = \text{Normal}(0.236, 0.390)\ \text{hours}^{-1}$$

Uncertainty about the mean and standard deviation were modelled using the Student-t and Chi-square distributions as appropriate for the Normal distribution assumption. A more technical analysis would have considered the batch nature of the samples, but there appeared to be no correlations in growth rate between clams in the same batch.

Assumptions

- The simple environment-independent growth model assumes that the same growth behaviour takes place post-retail as occurs from harvest to retail. This would be the case if the clams are held at the same temperature, pH, etc., from harvest to consumption, or if the growth rate is insensitive to environmental conditions over the range of conditions in the clams gut that are due to changes in the storage conditions.
- The exponential growth model does not allow for a tail-off of growth due to nutrition depletion, bacterial competition, etc. Since the range of growth in this model is less than one order of magnitude, this is unlikely to be a problem.
- All clam batches took the same time to get from harvest to retail and the method of transportation, etc., was representative of the production as a whole.

II–6.4 Number of pathogenic cfu in a random clam at the point of consumption

The time between retail and consumption (t_{rc}) was subjectively estimated, based on observations of individual behaviour, to be:

$$t_{rc} = \text{PERT}(0,1,25)\ \text{hours}$$

PERT distribution {PERT (a,b,c)} requires three parameters, namely minimum (a), most likely (b) and maximum (c). Based on observations at the harvesting site, we used a = 0, b = 1 and c = 25 hours to describe uncertainty (Vose, 2000: 71).

The number of *V. parahaemolyticus* cfu in a clam N_C was limited to a maximum of 10^9 and was thus estimated as:

$$N_C = MIN(N_r{*}Exp(k{*}t_{rc}), 1E9)$$

Limiting the maximum cfu per clam had very little effect on the simulated cfu distribution. The number of pathogenic cfu on a clam N_{CP} was then estimated as:

$$N_{CP} = N_C{*} (p_{tdh}{+} + p_{trh}{+})$$

Assumptions

- Pathogenic and non-pathogenic strains grow at the same rate.
- The time from retail to consumption is purely subjectively estimated. The risk estimate is sensitive to this estimate, so it would be useful to have more objective data on this variable.

II–6.5 The fraction of Bloody clam meals that are improperly cooked

One person in twenty surveyed had improperly cooked the clams. The uncertainty about the fraction of individuals that undercook clam was described as:

$$p_U = Beta(1+1, 20-1+1)$$

using the standard Bayesian estimate of a binomial probability (Beta(s+1, n-s+1), s=one, n=20 based on survey), with an uninformed prior (Vose, 2000).

Assumptions

- The twenty people surveyed are a representative random sample of Thai cooking practices.
- All Bloody clams are cooked boiled.
- The exposure path is through direct consumption of improperly cooked clams, not cross-contamination.
- Clams are either improperly cooked (all pathogens survive) or properly cooked (all pathogens are inactivated). No allowance is made for the possibility of inactivation of only a fraction of the pathogens.

II–6.6 Number of pathogenic V. parahaemolyticus *consumed in a random meal*

The number of Bloody clams N_C in a random meal was simulated by re-sampling from the reported consumption per meal. One sample was taken for the growth rate *k* and time from retail to consumption t_{rc} to better simulate that all clams in a meal will probably have had the same storage history. The number of pathogenic *V. parahamolyticus* cfu in a meal D was then simulated and added together to give a total dose of pathogenic *V.parahamolyticus* in a meal (D) as:

$$D = \sum_{i=1}^{n_c} N_{CP}(i)$$

where *Nc* is the number of clams consumed per meal, which was determined by interviews. This has the effect of making the spread of the distribution of cfu in a random meal much larger than if we had assumed that all clams were stored for independent amounts of time.

For each Bloody clam in the meal, the number of pathogenic *V. parahaemolyticus* was simulated and added together to give a total dose D. This resulted in a distribution of the sum of pathogenic *V. parahaemolyticus* in an improperly cooked meal as shown in Figure II–8.

Assumptions

- The people surveyed for their clam consumption patterns are a representative random sample of the Thai population. Comparison of people of different professional levels showed no difference in consumption patterns, but there may well be pockets of people who consume considerably more, for example, and such sub-population variation is not represented here.
- All clams in a meal come from the same batch, and have the same storage history.

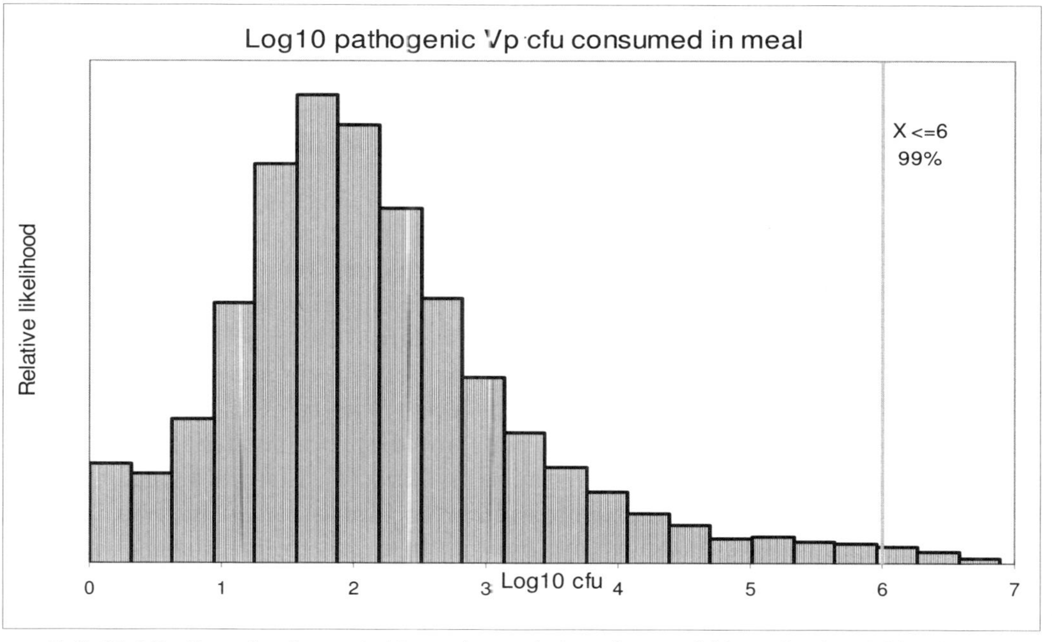

Figure II–8. Distribution of pathogenic *V. parahaemolyticus* (Log_{10} cfu) in a simulated Bloody clam model after improper cooking.

II–6.7 Individual annual consumption

The frequency of clams consumption per year and the number of Bloody clams consumed per meal were estimated based on interview data, and simulated by drawing a Bootstrap re-sample from the interview data that gave the number of clams in the simulated meal.

Assumption

- The people surveyed for their clam consumption patterns are a representative random sample of the Thai population. Comparison of people from different professional levels showed no difference in consumption patterns, but there may well be pockets of people who consume considerably more, for example, and such sub-population variation is not represented here

II–6.8 The probability of illness from a random simulated meal

The pathogenic load in a meal of Thai Bloody clams as modelled by this risk assessment remains with very high probability below 10^6 cfu, as shown in Figure II–8. For the FDA model, making the dose for the fitted curve, as they have done, ten times higher would bring the linear range up to 10^6 cfu, so the low dose approximation of Equation 7 in Appendix II–3 would be appropriate. In addition, the same linear relationship would hold if the P (illness | infection) was unrelated to consumed dose, i.e.

$$P(\text{illness} \mid \text{D}) \approx Dp\theta \propto D$$

where P is the probability of infection by a single cfu, and θ is P(illness | infection). In other words, whichever dose-response model we use, within the estimated exposure received by a Thai consuming a meal of Bloody clams, the probability of illness can be considered to be proportional to the number of cfu consumed. Extending that logic, the expected number of Thai *V. parahaemolyticus*–related illnesses in a year from Bloody clams will have the same constant of proportionality with the total number of *V. parahaemolyticus* cfu consumed in Bloody clams in a year. The latter statement can be made because the probability of illness for an individual is very small, so there is little likelihood of any overlap in exposure-to-illness events. In all of this argument there remains the difficulty of making direct comparisons between the susceptibility of Japanese volunteers in feeding trials, United States of America consumers of oysters and Thai consumers of Bloody clams.

The FDA Bootstrap estimates for α and β were used in the low-probability approximation of the Beta-Poisson dose-response model with high dose, as described in Appendix II–3. If P is small, P(illness) = 1 -(1+λ/β)^-α, and if dose is large, d = λ:

$$P(\text{illness}) = 1 - \left(1 + \frac{D}{\beta}\right)^{-\alpha}$$

As pointed out in the FDA report, this is likely to be a significant (by a factor of ten or more) over-estimation of the risk associated with a contaminated meal. Whilst it does not provide a realistic estimate of *V. parahaemolyticus* risk in Bloody clams, it does give an upper bound to that risk.

Assumptions

- Linear dose-response approximation holds.
- All cases of *V. parahaemolyticus* illness are sporadic.
- All cases are independent, so no immunity is acquired from exposure to pathogenic cfu.
- Probably unrealistically high *P*(illness) is being modelled as an upper bound.

II–6.9 Expected number of times a person will become ill in a year with V. parahaemolyticus from Bloody clam

P(illness) was calculated for the simulated meal and multiplied by the number of meals, N_M, a person would be expected to eat of this size (number of clams) in a year and by the probability that the clams would not be properly cooked, p_U. The mean of the values for this calculation taken over all iterations of the model is equivalent to an upper estimate of the expected number of times a person would get ill with *V. parahaemolyticus* from consuming Bloody clams.

Assumption

- All cases of *V. parahaemolyticus* illness come from Bloody clams.

II–6.10 Results of simulation model

The simulation model estimated that the upper approximation of the expected number of times a person would get ill with *V. parahaemolyticus* from consuming Bloody clam is 3.56×10^{-3}, or roughly a probability of 4 in 1000 per person per year.

The estimation is perhaps small compared with the perceived rate of illness, especially given the probable ten-fold over-estimate of the probability of illness given dose. This suggests that Bloody clam, or at least direct consumption of Bloody clam rather than cross-contamination of other uncooked foods, may not be the most important risk factor for *V. parahaemolyticus* in Thailand.

Some caution is needed in interpreting these results as they rely on a number of assumptions and relatively little data. A sensitivity analysis can be performed to determine which are the key random variables in the model driving the risk estimate.

Sensitivity analysis was undertaken as follows:

- Select an input distribution that is considered important in influencing the risk estimate (a quick Tornado chart will often help identify these variables).
- Select a number of cumulative percentiles to test: here 1%, 5%, 25%, 50%, 75%, 95% and 99% were used.
- Select one input distribution and replace it with its 1 percentile value (a practical minimum), then run the simulation and record some statistic of the output (here we have used the output mean).

- Repeat for all the other percentiles of that input.
- Then set that input back to being a distribution, and repeat the exercise with all the other significant input distributions.

The result is the spider plot of Figure II– 9. It shows, for example, that if the number of clams consumed is very low (at say the 1 percentile of the people surveyed), the risk is about 3×10^{-4}. The greater the vertical variation that an input parameter creates, the more the risk estimate is sensitive to that parameter's uncertainty or variability.

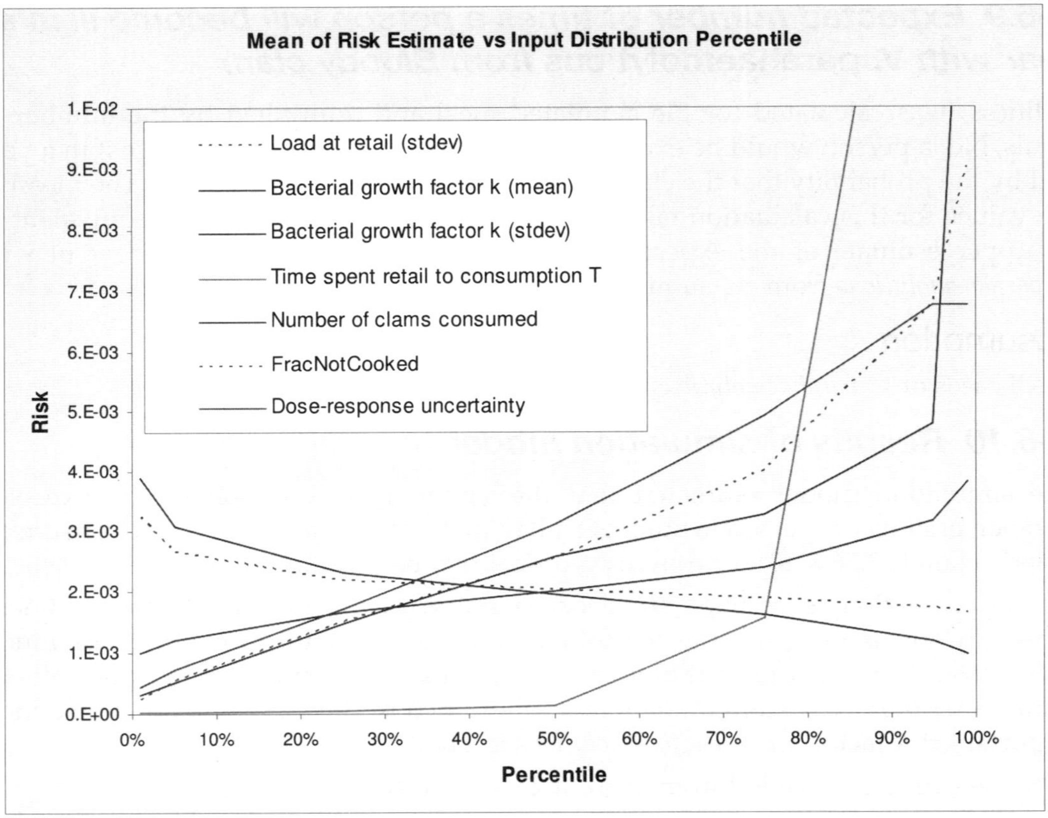

Figure II–9. Sensitivity analysis spider plot, where x-axis is percentile of uncertain parameter and y-axis is mean of the risk estimate (illnesses per person per year).

NOTES: Load at retail (stdev) = standard deviation of total *V. parahaemolyticus* concentration at the retail stage; Bacterial growth factor k (mean) = mean value of the growth rate of *V. parahaemolyticus*; Bacterial growth factor k (stdev) = standard deviation of the growth rate of *V. parahaemolyticus*; Time spent retail to consumption T = the maximum value for the time between retail and cooking at home; Number of clams consumed = the number of clams consumed at a meal derived from individuals who received interview; FracNotCooked = the fraction of individuals that undercook clam meals; Dose-response uncertainty = selection of parameter estimates for the Beta-Poisson dose-response model according to the likelihood of Bootstrap outcome by USFDA.

The spider plot shows that the mean of the risk estimate is most affected by the parameters t (time from retail to consumption) and the number of clams consumed in a year. Both these parameters are human behaviour variables; this suggests that our greatest ability to control *V. parahaemolyticus* illness may be in the education of people to handle the clams better. The next most influential parameters are the fraction of people who do not cook the clams properly (estimated from studying just twenty people) and the uncertainty in the dose-response model. The uncertainty in these two parameters could be reduced through further research.

In conclusion, the level of statistical uncertainty in the model's parameters means that the estimate from this model would change to perhaps as high as 0.01/person/year if some of the uncertain parameters turned out to be at their extremes, but would probably be no greater than that.

II–7. Fractional change model

Data are available on the growth of *V. parahaemolyticus* bacteria from harvest to retail and were analysed to provide the estimate of *k* in Section 4. This gave a change in the bacterial load with factor: $\frac{N_r}{N_h} = \exp(kt)$

where N_r = number of *V. parahaemolyticus* at retail stage and N_h = number of *V. parahaemolyticus* at harvest stage, *k* = Normal(0.236, 0.390) hours^{-1} and t = 5 hours from before, and where the Normal distribution describing *k* is modelling the variation in growth rate between harvest-retail samples. From the linear dose-response approximation we can conclude that if the clams were kept sufficiently chilled between harvest and retail to prevent any bacterial growth, the number of human illnesses from *V. parahaemolyticus* from Bloody clam would be reduced by a factor Exp(*kt*). In fact, we would want to model just the mean reduction in bacteria, not its variation from one clam to another, so the factor is:

$$\text{Reduction} = Exp(\bar{k}t)$$

where $\bar{k}$ is the estimated mean for *k*, and follows a Student-t distribution.

Figure II–10 shows the result of simulating this formula, giving a mean reduction of 3.5:

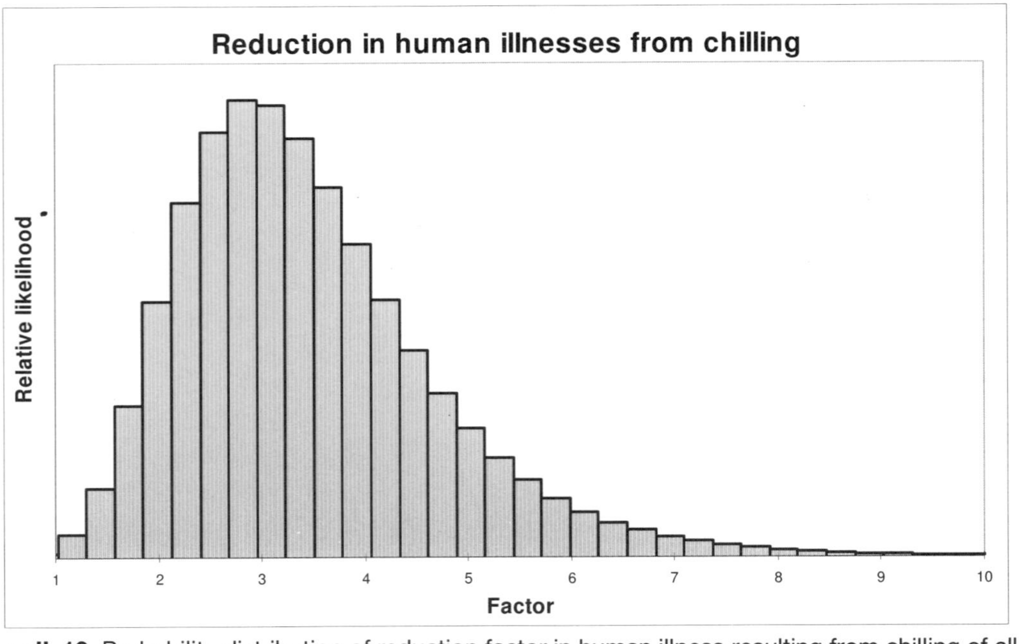

Figure II–10. Probability distribution of reduction factor in human illness resulting from chilling of all Bloody clam between harvest and retail.

II–8. Comparison between "farm-to-fork" and "fractional change" models

The fractional change model is appealing in its apparent simplicity, particularly in comparison with the more complicated farm-to-fork type models that are usually undertaken. A few important points need to be thoroughly evaluated when selecting between these model types.

A fractional change model essentially assumes a proportional relationship between the exposure unit being modelled and the resultant underlying human health risk. When that assumption is true we can state, for example, that halving the exposure (number of exposure events or average amount of exposure units in a random exposure event, or some combination of the two) will halve the human health impact. The beauty (and the danger) of a fractional change model is that one bypasses having to model all the intermediary steps, for which we may have insufficient data.

The unit of exposure could be a portion of contaminated food (assuming that food is the only exposure pathway of interest), as in the FDA-CVM risk assessment of Fluoroquinolone-resistant Campylobacter (FDA-CVM, 2001), in which case the model will only be able to investigate risk management actions that affect the number of contaminated food portions consumed. Such a model implicitly assumes that the probability distribution of bacteria for contaminated portions and all other elements of the system not deliberately altered by risk management actions remain substantively unchanged.

The unit of exposure could also be a viral or bacterial unit, as in this example. In which case we are assuming that the number of bacteria or virus particles in an exposure event remains within the linear range of the dose-response function, before and after any risk management actions, i.e. that the probability of illness is proportional to the amount of pathogen a person is exposed to for any random exposure.

Farm-to-fork models do not have these restrictions but face another, usually far greater, problem, that of insufficient data. A farm-to-fork approach attempts to model pathogen emergence, redistribution, growth and attenuation throughout the many exposure pathways, and then finishing with a dose-response model to predict whether the simulated exposure results in an adverse health event. The data needs can be enormous, and are—almost without exception—never completely met, even for the wealthiest countries. The models are usually huge, increasing the risk of error and making them difficult to update. There are also far more assumptions that need to be made, which introduce significant model uncertainty. For example, in producing a farm-to-fork model one must believe that all the important exposure pathways have been identified, but there is still a great deal of epidemiological debate for even the most common food pathogens concerning how much risk can be attributed to various activities resulting in exposure to that pathogen.

At the same time, farm-to-fork models attempt to specify as much of the process as possible, with the view to better understand the system. Where this is achievable, the model can potentially provide greater insight for risk managers.

A fractional change model requires that one can separately estimate the human health impact of exposures to the pathogen via the routes in question. One may be able to state that the human health impact can be halved, but without knowing what the rate was that is now halved, that information is of very limited value. Farm-to-fork models, in contrast, may be able to predict the amount of human health risk, from load, prevalence and dose-response data, and may be able to compare the risks from different pathways. Risk assessors sometimes try to 'calibrate' a farm-to-fork model to make it produce a level of risk that matches what has been observed. They do this by allowing one or more unknown model parameters to be altered so the final model matches reality. Although in some special cases this is mathematically valid, in general such an approach is fraught with danger: for example, this can reinforce a preconceived notion that a particular exposure pathway is causing all the observed illness. The parameter p in the dose-response function of the harvest-to-consumption model above was extremely uncertain and could have been set to produce a large number of illnesses, which would have stopped us from concluding that Bloody clam was not a major health concern. Ultimately, the choice of model structure should reflect the available data and the decision questions to which it is responding.

II–9. Conclusions and data gaps

This report has attempted to demonstrate the process of constructing a risk assessment model by first outlining what the problem is; what data are available, including how well we know the link between the pathogen–food type combination and the resultant human health impact; and then asking what models can be made with that available data to give decision-makers quantitative information about the benefits and costs (if any) of specific remedial actions.

A harvest to consumption risk assessment was conducted using one of the most popular seafood species in a tropical region of the world. In this case study, transportation time and temperature distributions of the environment in which clams were kept, prevalence and concentration of the pathogen at each stage of the production chain, and consumption patterns were modelled. This assessment was undertaken for a limited period and restricted to a single food item, and it is recognized that the sample size and sampling times might not be sufficiently large. Nevertheless, this project serves as a case study for initial data generation and risk assessment modelling in a developing country, with limited time, resources and quantitative data.

A second model, capitalizing on the linear dose-response approximation that could be made, considered the fractional change in human illnesses that would occur with a measure that were to control the growth of bacteria.

II–10. Key findings

1. This study estimated that only a few people per 10 000 population per year acquire *V. parahaemolyticus* infection as a result of consuming boiled Bloody clam. Therefore the present risk estimate does not appear to support the common perception of the population of Hat Yai City that Bloody clam is a major cause of diarrhoeal illness, including *V. parahaemolyticus* illnesses. These results are compatible with the following observations:

- The results of laboratory research detecting the virulent strains in bivalves sampled in 2000 and 2001 showed that Bloody clam might not be the most common shellfish source of virulent strains of *V. parahaemolyticus*, which were similar to those isolated from patients, although there was still shellfish sampling bias (V. Vuddhakul, pers. comm.). This might indicate the need for future risk assessment of *V. parahaemolyticus* in two other shellfish species.
- Most people (19/20) seem to cook Bloody clam to a temperature that would destroy *V. parahaemolyticus* organisms.

2. At the same time, this study may underestimate the risk of Bloody clam-associated *V. parahaemolyticus* illness for a number of reasons that could not be addressed in this initial phase of data generation and might be targeted as data gaps that could be filled in the future, namely:

- Cooking methods other than boiling were not modelled. It is known that methods such as grilling and frying are used in Hat Yai City, and they are less likely than boiling to eliminate the pathogen. Additionally, some people may purchase local salads (called *pla* and *yam*) containing (improperly) cooked Bloody clam from the delicatessen, and leave them at ambient temperature for some time before consuming at home. These alternative methods of preparing and consuming Bloody clam might be a source of *V. parahaemolyticus* infections in Thailand.
- Other factors that might influence the frequency of contamination from harvest to consumption need to be considered and modelled, such as the possibility that virulent strains are more resistant to heating than non-virulent strains.

II–11. Recommendations

It is recommended that a case-control study be conducted using patients in Hat Yai City with microbiologically confirmed *V. parahaemolyticus* infections. This would contribute greatly to the information base on incidence of the disease and provide more causal evidence for *V. parahaemolyticus* infections, and attribute the risk to the various exposure paths, both food- and environment-related. It might also provide more realistic evidence of behaviour that reduces or increases the risk of *V. parahaemolyticus* illness.

Bacterial data sets for Bloody clam throughout the food chain should be collected, using state-of-the-art techniques for detecting virulent strains. More detailed data on peoples' behaviour in relation to harvesting, storage, cooking and consumption patterns need to be collected.

Together, such information would allow better evaluation of the management options discussed.

II–12. References for Part II – Bloody clam

Anon[ymous]. 2001a. Hazard identification, exposure assessment and hazard characterization of *Campylobacter* spp. in broiler chickens and *Vibrio* spp. in seafood. Ad hoc expert consultations on risk assessment of microbiological hazards in foods. Joint FAO/WHO Expert Consultation on Risk Assessment of Microbiological Hazards in Foods. WHO Headquarters, Geneva, Switzerland, 23–27 July 2001. FAO, Rome/WHO, Geneva. See ftp://ftp.fao.org/es/esn/jemra/CV01_en.pdf

Anon. 2001b. Draft Risk Assessment on the Public Health Impact of *Vibrio parahaemolyticus* in raw molluscan shellfish. Center for Food Safety and Applied Nutrition, U.S. FDA, Washington, DC. See http://www.cfsan.fda.gov/~dms/vprisk.html

Anon. 2001c. Annual epidemiological surveillance report. Division of Epidemiology, Office of Permanent Secretary for Public Health, Ministry of Public Health, Nonthaburi, Thailand.

Center for Food Safety and Applied Nutrition, Food and Drug Administration, U.S. Department of Health and Human Services. 2000 Draft Risk Assessment on the Public Health Impact of *Vibrio parahaemolyticus* in Raw Molluscan Shellfish. December, 2000.

Chowdhury, A., Ishibashi, M., Nakano, Y. & Nishibuchi, M. 2001. Prevalence of *Vibrio parahaemolyticus* in imported fish and shellfish. *Jpn. J. Soc. Bacteriol.*, 56(1): 323; 56(1): 215. Published in Japanese.

Cook, D.W., Bowers, J.C. & DePaola, A. 2002. Density of total and pathogenic (*tdh+*) *Vibrio parahaemolyticus* in Atlantic and Gulf Coast molluscan shellfish at harvest. *Journal of Food Protection*, 65: 1873–1880.

Daniels, N., MacKinnon, L., Bishop, R., Altekruse, S., Ray, B., Hammond, R., Thompson, S., Wilson, S., Bean, N., Griffin, P. & Slutsker, L. 2000a. *Vibrio parahaemolyticus* infections in the United States, 1973–1998. *Journal of Infectious Disease*, 181: 1661–1666.

Daniels, N.A., Ray, B., Easton, A., Marano, N., Kahn, E., McShan, A.L., Del Rosario, L., Baldwin, T., Kingsley, M.A., Puhr, N.D., Wells, J.G. & Angulo, F.J. 2000b. Emergence of a new *Vibrio parahaemolyticus* serotype in raw oysters. *Journal of the American Medical Association*, 284: 1541–1545.

Dileep, V., Kumar, H.S., Kumar, Y., Nishibuchi, M., Karunasagar, Indrani & Karunasagar, Iddya. 2003. Application of polymerase chain reaction for detection of *Vibrio parahaemolyticus* associated with tropical seafoods and coastal environment. *Letters in Applied Microbiology*, 36: 423–427.

Echeverria, P., Pitarangsi, C., Eampokalap, B., Vibulbandhitkit, S., Boonthai, P., and Rowe, B. 1983. A longitudinal study of the prevalence of bacterial enteric pathogens among adults with diarrhea in Bangkok, Thailand. *Diagn. Microbiol. Infect. Dis.*, 1(3): 193–204.

Fang, S.W., Huang, W.W. & Chen, L.H. 1987. Contamination of seafood by *Vibrio parahaemolyticus* in Taiwan. *Zhonghua Min Guo Wei Sheng Wu Ji Mian Yi Xue Za Zhi*, 20(2): 140–147.

FAO/WHO. 2003. Hazard Characterization for Pathogens in Food and Water - Guidelines. *Microbiological Risk Assessments Series*, No. 3. Available at: http://www.who.int/foodsafety/publications/micro/pathogen/en/

FDA [US Food and Drug Administration]. 2005. Quantitative Risk Assessment on the Public Health Impact of Pathogenic *Vibrio parahaemolyticus* in Raw Oysters. U.S. Department of Health and Human Services, U.S. Food and Drug Administration. See: http://www.fda.gov/Food/ScienceResearch/ResearchAreas/RiskAssessmentSafetyAssessment/ucm185746.htm

FDA-CVM [Center for Veterinary Medicine]. 2001. Risk Assessment on the Human Health Impact of Fluoroquinolone Resistant Campylobacter Associated with the Consumption of Chicken. See: http://www.fda.gov/cvm/antimicrobial/antimicrobial.html

Fujino, T., Sakaguchi, G., Sakazaki, R. & Takeda, Y. (editors). 1974. International Symposium on *Vibrio parahaemolyticus*. Tokyo: Saikon Publishing Co.

Haas, C.N., Rose, J.B. & Gerba, C.P. 1999, *Quantitative Microbial Risk Assessment*. John Wiley & Sons, New York, USA. See pp. 266–268.

ICMSF [International Commission on Microbiological Specifications for Foods]. 1996. *Vibrio parahaemolyticus*. Micro-organisms in Foods 5 – Characteristics of Microbial Pathogens. Blackie Academic & Professional, London. pp. 426–435

Khuharat, S. 1998. Food poisoning among participants attending intensive course, University A, Nonthaburi, Thailand, September 1996. *Weekly Epidemiology Surveillance Report*, 29: 477–493.

Laohaprertthisan, V., Chowdhury, A., Kongmuang, U., Kalnauwakul, S., Ishibashi, M., Matsumoto, C. & Nishibuchi, M. 2003. Prevalence and serodiversity of the pandemic clone among the clinical strains of *Vibrio parahaemolyticus* isolated in southern Thailand. *Epidemiology and Infection*, 130(3): 395–406.

Matsumoto, C., Okuda, J., Ishibashi, M., Iwanaga, M., Garg, P., Rammamurthy, T., Wong, H.-C., DePaola, A., Kim, Y.B., Albert, M.J. & Nishibuchi, M. 2000. Pandemic spread of an O3:K6 clone of *Vibrio parahaemolyticus* and emergence of related strains evidenced by arbitrarily primed PCR and *toxRS* sequence analysis. *Journal of Clinical Microbiology*, 38: 578–585.

Matte, G.R., Matte, M.H., Sato, M.I., Sanchez, P.S., Rivera, I.G. & Martins, M.T. 1994. Potentially pathogenic vibrios associated with mussels from a tropical region on the Atlantic coast of Brazil. *Journal of Applied Bacteriology*, 77(3): 281–287.

Mead, P.S., Slutsker, L., Dietz, V., McCaig, L.F., Bresee, J.S., Shapiro, C., Griffin, P.M. & Tauxe, R.V. 1999. Food-related illness and death in the United States. *Emerging Infectious Disease,* 5(5): 607–625.

Ministry of Health, Labour and Welfare. 1999–2001. National statistics of foodborne illness in Japan. Ministry of Health, Labour and Welfare, Japan.

Miyamoto, Y., Kato, T., Obara, Y., Akiyama, S., Takizawa, K. & Yamai, S. 1969. *In vitro* hemolytic characteristic of *Vibrio parahaemolyticus*: its close correlation with human pathogenicity. *Journal of Bacteriology,* 100: 1147–1149.

Nishibuchi, M. 2002. Report on the WHO project regarding "*Vibrio* Risk Assessment". Project: Food Safety, Obligation no. HQ/02/419664

Okuda, J., Ishibashi, M., Hayakawa, E., Nishino, T., Takeda, Y.,Mukhopadhyay, A., Garg, S., Bhattacharya, S.K., Nair, G.B. & Nishibuchi, M. 1997. Emergence of a unique O3:K6 clone of *Vibrio parahaemolyticus* in Calcutta, India, and isolation of strains from the same clonal group from Southeast Asian travelers arriving in Japan. *Journal of Clinical Microbiology,* 35: 3150–3155.

Sakazaki, R., Tamura, K., Kato, T., Obara, Y., Yamai, S. & Hobo, K. 1968. Studies of the enteropathogenic, facultatively halophilic bacteria, *Vibrio parahaemolyticus.* III. Enteropathogenicity. *Japanese Journal of Medical Science Biology,* 21: 325–331.

Sumner, J., DePaola, A., Osaka, K., Karunasager, I., Walderhaug, M., Bowers, J. 2001. Hazard identification, exposure assessment and hazard characterization of *Vibrio* spp. in seafood. FAO/WHO MRA 01/03/04.

Suthienkul, O. 2000. Situation of food microbial and public health. *Thai Journal of Epidemiology,* 8: 134-151.

Tangkranakul, W., Tharmaphornpilas, P., Datapon, D. & Sutantayawalee, S. 2000. Food poisoning outbreak from contaminated fish-balls. *Journal of the Medical Association of Thailand,* 83: 1289–1295.

USFDA. 2005. Quantitative Risk Assessment on the Public Health Impact of Pathogenic *Vibrio parahaemolyticus* in Raw Oysters. Center for Food Safety and Applied Nutrition, U.S. FDA, Washington, DC. http://www.cfsan.fda.gov/~dms/vpra-toc.html

Vose, D. 2000. *Risk Analysis: a quantitative guide.* Wiley, Chichester, UK.

Vuddhakul, V., Chowdhury, A., Laohaprertthisan, V., Pungrasamee, P., Patararungrong, N., Thianmontri, P., Ishibashi, M., Matsumoto, C. & Nishibuchi, M. 2000. Isolation of a pandemic O3:K6 clone of a *Vibrio parahaemolyticus* strain from environmental and clinical sources in Thailand. *Applied and Environmental Microbiology,* 66(6): 2685–2689.

Vuddhakul, V., Soboon, S., Sunghiran, W., Kaewpiboon, S., Chowdhury, A., Ishibashi, M., Nakaguchi, Y. & Nishibuchi, M. 2006. Distribution of virulent and pandemic strains of *Vibrio parahaemolyticus* in three molluscan shellfish species (*Meretrix meretrix, Perna viridis,* and *Anadara granosa*) and their association with foodborne disease in southern Thailand. *Journal of Food Protection,* 69(11): 2615–2620.

Wagatsuma, S. 1974. Ecological studies on Kanagawa phenomenon positive strains of *Vibrio parahaemolyticus.* International symposium on *Vibrio parahaemolyticus.* Saikon Publishing Co.,Tokyo. pp. 91–96.

Wong, H.C., Ting, S.H. & Shieh, W.R. 1992. Incidence of toxigenic vibrios in foods available in Taiwan. *Journal of Applied Bacteriology,* 73(3): 197-202.

Wong, H.C., Chen, M.C., Liu, S.H. & Liu, D.P. 1999. Incidence of highly genetically diversified *Vibrio parahaemolyticus* in seafood imported from Asian countries. *International Journal of Food Microbiology,* 52(3): 181–188.

Yamamoto, A., Iwahori, J.,Vuddhakul, V., Jaroenjiratrakul, W., Vose, D., Osaka, K., Shigematsu, M., Toyofuku, H., Yamamoto, Y., Nishibuchi, M. & Kasuga, F. 2008. Quantitative modeling for risk assessment of *Vibrio parahaemolyticus* in bloody clams in southern Thailand. *International Journal of Food Microbiology*, 124: 70–78.

Appendix II-1
Methods used for isolation and characterization of *V. parahaemolyticus* strains from clinical and seafood specimens for hazard identification

Clinical strains

Stool or rectal swab samples were collected from patients with diarrhoea. The stool specimens and rectal swabs were inoculated into Stuart's transport medium and maintained until bacteriological examination. The samples were plated directly onto thiosulfate-citrate-bile salt-sucrose agar (TCBS agar). The green colonies detected on the medium were examined by standard biochemical tests for identification of *V. parahaemolyticus*. Strains screened by the biochemical tests were examined by the PCR method for detection of the *V. parahaemolyticus*-specific *toxR* gene sequence (Kim et al., 1999). The strain that gave a positive result in this PCR test was identified as *V. parahaemolyticus*.

Seafood strains

Fresh seafood samples were purchased in a fresh market. The seafood samples were transferred to the laboratory at room temperature and examined within 2 hours after sample collection. Shellfish were shucked aseptically and subjected to examination. Ten to 30 g of each seafood sample were inoculated into 100 mL of alkaline peptone water at pH 9.2, mixed using a stomacher for 30 sec, and the mixture allowed to settle for a few minutes. Only the supernatant was transferred into a sterile flask and incubated without shaking at 37°C for 7 h. After incubation of the seafood in alkaline peptone water, 2 mL of the broth culture was mixed with 20 mL of the anti-K antiserum or antisera (explained in detail below) and incubated for 20 min at room temperature with gentle mixing. Bacterial cells were harvested by centrifugation, washed with 2 mL of PBS and the cells were suspended in 1 mL of PBS. Ten mL of the anti-rabbit IgG-coated Dynabead suspension was added to the cell suspension, and incubated for 20 min with intermittent mixing. The immunomagnetic beads were collected, washed twice with 2 mL aliquots of PBS, suspended in 10 mL of PBS, and then the entire bead suspension was streaked onto a TCBS agar plate. After 18 to 24 h of incubation at 37°C, five green colonies were selected and examined for biochemical characteristics to screen for *V. parahaemolyticus*. The strains showing motility in nutrient agar-based semi-solid medium and positive results in oxidase, lysine decarboxylase and indole tests were selected. The strains thus screened were examined by the PCR method for detection of the *V. parahaemolyticus*-specific *toxR* gene sequence. Strains that gave a positive result in this PCR test were identified as *V. parahaemolyticus*.

O:K serovar

The test strains were grown and their O:K serovars were determined by an agglutination test using specific antisera, as described previously by Suthienkul et al. (1995).

PCR

The presence or absence of the *tdh* and *trh* genes in test strains was determined by the PCR method using primers D3 and D5 for the *tdh* gene and primers R2 and R6 for the *trh* gene, as described previously by Tada et al. (1992). GS-PCR to identify the strain belonging to the pandemic clone was performed as described previously by Matsumoto et al. (2000), with a minor modification. The method to prepare the PCR template was modified to replace the broth culture with the bacterial cells suspended in saline before the boiling step, as follows. The test strain was grown in Luria-Bertani (LB) broth medium containing 1% NaCl at 37°C with shaking (160 rpm) overnight. One mL of the culture was centrifuged (14 000 rpm) on a tabletop centrifuge (Centrifuge 5415C; Eppendorf, Hamburg, Germany) at room temperature. The supernatant was discarded and the pelleted bacterial cells were suspended in 1 mL of saline (0.85% NaCl) and then boiled for 10 min and transferred onto ice immediately. The supernatant was then obtained by centrifugation at room temperature. The supernatant was diluted 10-fold with distilled water and used as the template solution for PCR.

References

Kim, Y.B., Matsumoto, C., Takahashi, N., Hashimoto, S. & Nishibuchi, M. 1999. Identification of *Vibrio parahaemolyticus* at the species level by PCR targeted to the *toxR* gene. *Journal of Clinical Microbiology*, 37: 1173–1177.

Matsumoto, C., Okuda, J., Ishibashi, M., Iwanaga, M., Garg, P., Rammamurthy, T., Wong, H.-C., Depaola, A., Kim, Y.B., Albert, M.J. & Nishibuchi, M. 2000. Pandemic spread of an O3:K6 clone of *Vibrio parahaemolyticus* and emergence of related strains evidenced by arbitrarily primed PCR and *toxRS* sequence analyses. *Journal of Clinical Microbiology*, 38: 578–585

Suthienkul, O., Ishibashi, M., Iida, T., Nettip, N., Supavej, S., Eampokalap, B., Makino, M. & Honda, T. 1995. Urease production correlates with possession of the *trh* gene in *Vibrio parahaemolyticus* strains isolated in Thailand. *Journal of Infectoius Disease*, 172: 1405–1408.

Tada, J., Ohashi, T., Nishimura, N., Shirasaki, Y., Ozaki, H., Fukushima, S., Takano, J., Nishibuchi, M. & Takeda, Y. 1992. Detection of the thermostable direct hemolysin gene (*tdh*) and the thermostable direct hemolysin-related hemolysin gene (*trh*) of *Vibrio parahaemolyticus* by polymerase chain reaction. *Molecular and Cellular Probes*, 6: 477–487.

Appendix II-2
Methods used for exposure assessment

Examination of bivalve samples

The bivalve samples obtained as described in the section on *Measurement of time and temperature of the seafood*, below, were examined by the following procedure:

1. Inoculate the bivalve sample (one or more animals) into 9 volumes of alkaline peptone water (APW), homogenize, and allow the sample to settle in APW (hereinafter called APW enrichment broth).
2. Inoculate 10 mL, 1 mL, 0.1 mL, 0.01 mL and 0.001 mL equivalent of the APW enrichment broth into a sterile tube containing 10 mL APW in triplicate (3 tube MPN for 5 dilutions); the last two dilutions prepared by dilution starting from the tube inoculated with 0.1 mL APW enrichment broth. Incubate the inoculated tubes at 35 to 37°C overnight.
3. Transfer a 1-mL portion of each of the broth cultures to a sterile tube containing 10 mL salt polymixin broth (SPB) and incubate at 35 to 37°C overnight.
4. Transfer a 1-mL portion of each of the broth cultures to a sterile tube containing 10 mL SPB and incubate at 35 to 37°C for 6 hours.
5. (a) **Direct PCR method**: Take a 1-mL portion of each SPB culture and prepare the boiled supernatant for PCR assay, and perform PCR for detection of the *V. parahaemolyticus-toxR* gene, the *tdh* gene and the *trh* gene, as described above.

   (b) **Culture method**: Inoculate a loopful of SPB culture from each of the broth culture onto CHROMagar Vibrio (Hara-Kudo et al., 2001) and incubate at 35 to 37°C overnight. Pick up two typical violet colonies from CHROMagar and test the isolated strain(s) by the PCR method for the *V. parahaemolyticus-toxR* gene to confirm that the picked colonies are *V. parahaemolyticus*, and by the PCR methods to detect the *tdh* and *trh* genes. Judge positive if at least one colony gives positive result in each of the PCRs for the *V. parahaemolyticus-toxR*, *tdh*, and *trh* genes.

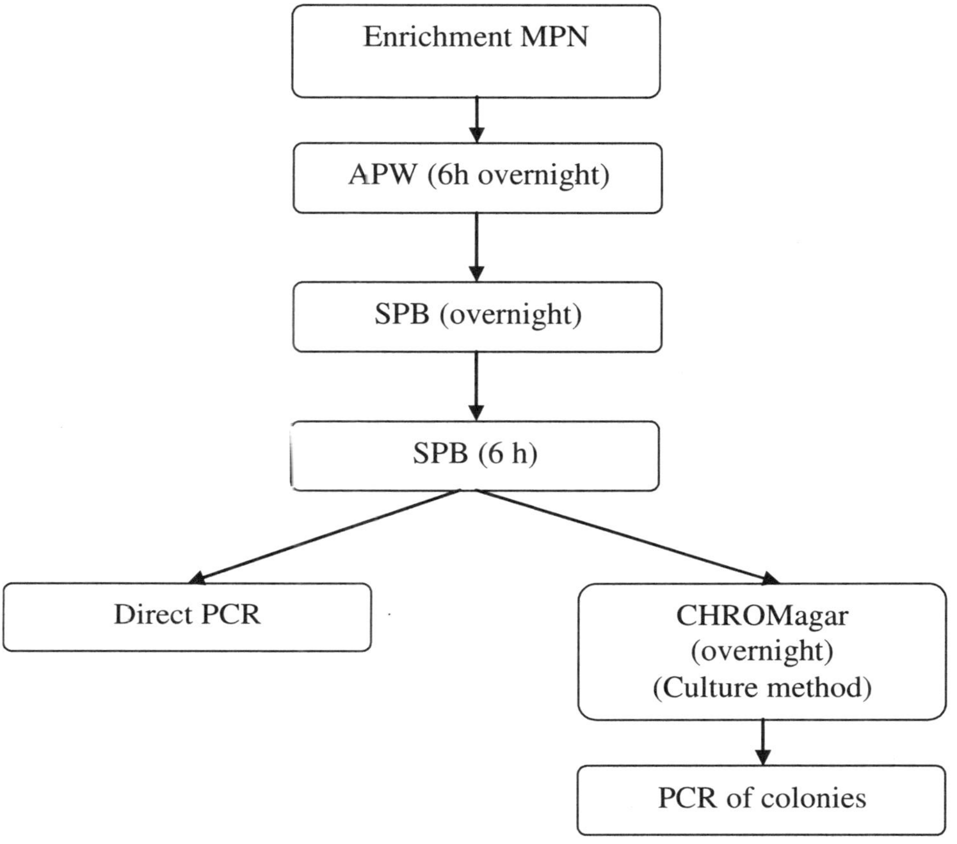

Measurement of time and temperature of the seafood

One lot of the bivalves was obtained soon after landing from the boat at a harvest site. After the initial examination (inoculation into APW) of a suitable number of the bivalves, the rest of the bivalves were transported and bivalve samples were picked so that the actual retail and consumption conditions were simulated. Retail means selling bivalves at the fresh market, and consumption means cooking immediately followed by consumption at home. The picked samples were examined as described above in *Examination of bivalve samples*. The following parameters were monitored during the sampling:

- temperature inside the bivalve at each sampling point (harvest, retail and consumption); and
- transportation time and air temperature between each sampling point.

Reference

Hara-Kudo, Y., Nishina, T., Nakagawa, H., Konuma, H., Hasagawa, J. & Kumagai, S. 2001. Improved method for detection of *Vibrio parahaemolyticus*. *Applied and Environmental Microbiology*, 67(12): 5819–5823.

Appendix II-3 Technical explanation of Beta-Poisson dose-response model and its approximations

The basic probability model

The Beta-Poisson model is potentially well suited to feeding trial data because it is derived from the following principles:

If a person (the 'host') receives a dose of D viable pathogenic organisms and if each organism has the same, *independent* chance of infecting the host p, then the probability of infection is given by:

$$P(\text{infection}) = 1 - (1 - p)^D \qquad \text{Equation 1}$$

i.e. it is one minus the probability that none of the pathogen cause infection.

In a feeding trial it is not possible to count the actual dose D given to a person without affecting the pathogen somehow. However, through sequential dilution methods one can control the concentration of pathogen in a well-mixed volume of liquid and take a sample of that liquid to give as the dose. In this situation, the concentration λ, ie the expected number of pathogen, in that volume of liquid is known, and the actual dose is given by a Poisson distribution:

$$D = \text{Poisson}(\lambda) \qquad \text{Equation 2}$$

We can also recognise that inter-individual variations between batches of pathogen, or between hosts, will affect the probability p. A simple, perhaps simplistic, way of representing this inter-individual variation of interaction is to model p as a Beta distributed variable:

$$p = \text{Beta}(\alpha, \beta) \qquad \text{Equation 3}$$

The Beta distribution can take many shapes. At the expense of estimating just one extra parameter from the data (we would already have had to estimate p if we considered it fixed), it is a convenient tool to represent possible patterns of inter-individual variability of p.

[One should bear in mind that a feeding trial experiment will usually use just the one colony of pathogen, for which the virulence is probably fixed, so any feeding trial using just one original colony to produce all doses will not display the pathogen component of the inter-individual variability for p. Similarly, data for several feeding trials will only show variability from one experiment to another, not one administered dose to another.]

Putting together Equations 1, 2 and 3 we get:

$$P(\text{infection}) = 1 - {}_1F_1(\alpha, \alpha+\beta, -\lambda)$$

where $_1F_1$ is the Kummer confluent hypergeometric function – see, for example, Abramovitz M, Stegun IA (1984) *Pocketbook of mathematical functions.* This function is very complicated and difficult to evaluate, and we therefore look for approximations.

Approximation when p is small

A nice approximation when p is small (ie when $\alpha << \beta$) is:

$$\text{P(infection)} = 1 - \left(1 + \frac{\lambda}{\beta}\right)^{-\alpha} \qquad \text{Equation 4}$$

This is the most common function used and the one that the USFDA fitted to *V. parahaemolyticus* feeding trial data. The simplifying assumption that p is small is valid for *V. parahaemolyticus*, but Teunis and Havelaar (2000) have also shown that confidence intervals calculated on this approximate model can produce nonsensical answers.

Approximation when dose is large

A Poisson(λ) distribution has a mean λ and a standard deviation $\sqrt{\lambda}$. A Poisson distribution with large λ (>100 or so) also approximates a Normal distribution very well. Thus, for large values of λ, a Poisson(λ) variable has a coefficient of variation (standard deviation/mean) of 1/sqrt(λ). For example, for a dose with mean 10^6, the coefficient of variation is 10^{-3}, meaning that the variable varies relatively only minutely about its mean. At such high values, it is valid to assume that the dose is fixed, ie :

$$D = \lambda \text{ when } \lambda \text{ is large} \qquad \text{Equation 5}$$

Combining Equations 1, 3 and 5 gives the Beta-Binomial dose-response model, which is both an approximation to the Beta-Poisson dose-response model as just explained, and the appropriate model in a simulation model that generates values of actual number of pathogen ingested (ie not means):

$$\text{P(infection)} \approx 1 - \frac{\Gamma(\alpha+\beta)\Gamma(\beta+\lambda)}{\Gamma(\beta)\Gamma(\alpha+\beta+\lambda)} \qquad \text{Equation 6}$$

where $\Gamma(\bullet)$ is a gamma function.

Fixed probability p approximation

Where p is not considered to vary greatly between hosts or between pathogen colonies, the mean number of infections is λp, the probability of no infections is $e^{-\lambda p}$ and so the probability of at least one infection of the host is:

$$\text{P(infection)} = 1 - e^{-\lambda p}$$

Low dose approximation when p is small

When the probability p and the dose D are sufficiently small (pD < 0.1 approximately) the simplest possible approximation is reasonably accurate:

$$\text{P(infection)} \approx pD \qquad \text{Equation 7}$$

Probability of infection v Probability of illness

The Beta-Poisson model is frequently used to model the probability of infection, particularly for fitting to feeding trial data where the administered dose is known to be Poisson distributed, and where one usually obtains measures of infection. One problem lies in the definition of infection, and matching that to a physical observation. A usual method for determining infection is to observe the persistent expression of pathogen in the host's faeces, for example.

Illness is a clinical symptom and depends on the pathogen, but is usually measured as diarrhoeal illness with a specified minimum threshold of volume.

The greater contention for the risk modeller comes in modelling the probability of transfer from infection to illness, i.e. P(illness | infection). There is considerable debate as to whether this is a function of dose. Many scientists argue that once infection has been established the progress of disease is independent of how many bacteria were ingested. It that case, the probability of transition is constant for an individual:

$$P(\text{illness} \mid \text{infection}) = q$$

If the probability is strongly variable between individuals, and/or between pathogen colonies or strains, then we might consider using another Beta distribution to model q:

$$q = \text{Beta}(\phi, \theta) \text{ which has a mean } \phi/(\phi+\theta)$$

In either case, from an epidemiological perspective, we would expect to see a stable proportion of the population q that had become infected then go on to show signs of illness.

Reference

Teunis, P.F. & Havelaar, A.H. 2000. The Beta Poisson dose-response model is not a single-hit model. *Risk Analysis*, 20(4): 513–520.

Part III

Microbiological risk assessment of *Vibrio parahaemolyticus* in finfish

III–1. Introduction

Vibrio parahaemolyticus infections associated with raw and undercooked seafood other than oyster have a considerable impact in some Asian countries (FAO/WHO, 2002). However, an initial review of *V. parahaemolyticus* in finfish indicated a lack of relevant quantitative data with which to pursue a quantitative risk assessment on this pathogen–commodity combination. Nevertheless, during a subsequent expert review, it was agreed to continue the risk assessment of raw and undercooked finfish because some of the information presented at the Joint FAO/WHO Expert Consultation in 2002 in Bangkok (FAO/WHO, 2002) was considered to be useful to countries in implementing a risk assessment on finfish. It was decided to focus the risk assessment on a single appropriate finfish species in order to progress development of a model. Expansion of such a model to include other species and the additional complications of undercooked seafood could then be undertaken later if the need and appropriate resources were identified. Horse mackerel (*Trachurus japonicus*) was selected as the target finfish of this risk assessment because it is one of the most popular finfish harvested and consumed in many countries. After selecting the target finfish, additional quantitative data were collected through extensive literature reviews and by working with experts in this field to generate new data sets of several relevant scenarios, e.g. simulating washing effects by immersing the experimentally contaminated fish in clean seawater.

III–2. Scope

The purpose of the present risk assessment is to estimate the risk of *V. parahaemolyticus* infection associated with the consumption of raw horse mackerel in Japan, where large amounts of raw fish are consumed and relevant data are more easily accessible than in other areas, and to estimate the risk reduction from washing horse mackerel with disinfected seawater or potable water after harvest or during preparation.

III–3. Hazard identification

Increasing the occasions of eating raw fish and shellfish has also increased the possibility of *V. parahaemolyticus* infection. While consumption of raw oysters has been a major cause of *V. parahaemolyticus* infections in several countries, such as the United States of America (Anon., 1998b, 1999), seafood other than oysters has been considered an important vehicle in some countries. *V. parahaemolyticus* infections associated with oyster consumption accounted for less than 5% of all [reported] outbreaks due to *V. parahaemolyticus* in Japan (Anon., 2000a). Factors that might contribute to the low rate of *V. parahaemolyticus* illness associated with oysters in

Japan include: (i) consumption of larger amounts of raw seafood other than oyster; (ii) the implementation of microbiological criteria for oyster consumed raw; (iii) not harvesting oyster for raw consumption purposes during the summer months because the concentrations of *V. parahaemolyticus* in oysters exceed the microbiological criteria; and (iv) strict control of oysters from harvest to retail from October to April.

Outbreaks due to *V. parahaemolyticus* associated with fish and shellfish other than oysters have been reported in several countries. Sporadic infections and outbreaks of *V. parahaemolyticus* associated with clams, molluscan shellfish, crayfish, lobster and shrimp have been reported in the United States of America (Daniels et al., 2000; Bean et al., 1998). There was also an epidemiological case-control study in Guam Island, where seafood consumption was thought to be important in *V. parahaemolyticus* infection (Haddock and Cabanero, 1994). Fish-balls were` associated with an outbreak of food poisoning due to *V. parahaemolyticus* in Thailand (Tangkanakul et al., 2000). Fish, shellfish and raw oyster were incriminated in a *V. parahaemolyticus* outbreak in Spain (Molero et al., 1989). During 2003 and during late September 2004, more than 1230 cases of gastroenteritis reported in north-western Mexico were attributed to the consumption of raw or undercooked shrimp (Cabanillas-Beltran et al., 2006). A serious outbreak affecting 44 patients associated with consumption of shrimps imported from Asia occurred in France in 1997 (Robert-Pillot et al., 2004). A more recent outbreak involving 80 illnesses of *V. parahaemolyticus* infection associated with consumption of boiled crab was reported in Spain in July 2004 (Martinez-Urtaza et al., 2005).

In Japan, typically 500–800 *V. parahaemolyticus* outbreaks affecting around 10 000 people are reported annually. The annual reports from prefectural public health institutes and local health centres on isolation of *V. parahaemolyticus* from human sources gave values of 922, 1516, 2507, 1904, 721, 651, 428, 304, 582, 310, 209 and 199 from 1996 to 2007, showing a tendency to increase from 1996, with a peak in 1998, and then a decrease (Otomo and Yatsuyanagi, 2003; IASR, no date). Implicated foods include *sashimi* (pieces of raw fish fillet; responsible for 26% of outbreaks), followed by *sushi* (vinegary rice ball with pieces of raw fish fillet; 23%), shellfish (16%) and cooked seafood (12%) (Anon., 2000a). Most outbreaks were considered to be associated with consuming seafood, but the causative food item was rarely identified. Among outbreaks in which causative food items could be traced in 1999, at least three outbreaks were associated with horse mackerel (Table III–1).

Table III–1. *V. parahaemolyticus* foodborne outbreaks in 1999 in Japan where the implicated foods were identified during outbreak investigations.

Location or source of seafood	Type of seafood	Serotype
Hokkaido	Boiled crab	O3:K6
Tottori Prefecture	Fresh fish	O3:K6
Wakayama Prefecture	Horse mackerel	Various types
A, Nagasaki Prefecture	Horse mackerel	ND
C, Nagasaki Prefecture	Horse mackerel	O4:K55
A, Nagasaki Prefecture	Jack-knife clam	O4:K8
Kumamoto Prefecture	Mysids	O3:K6, O11K
B, Nagasaki Prefecture	Olive shell	O3:K6
Republic of Korea	Pen shells	O3:K6, O4:K13
Democratic People's Republic of Korea	Pen shells	O3:K6 and others
Chile	Pickled turban shell	O3:K6, OUT:KUT
Ishikawa Prefecture	Rock oyster	ND
D, Nagasaki Prefecture	Sardines	O3:K6
Niigata Prefecture	*Sashimi*	O3:K6
Republic of Korea	*Sashimi*	O3:K6 and others
City A, Hokkaido	Scallops	O3:K6 and others
City B, Hokkaido	Scallops	O3:K6
Iwate Prefecture	Sea squirt	O3:K6
Iwate Prefecture	Sea squirt	O3:K6
City B, Hokkaido	Sea urchin	ND
Aomori Prefecture	Sea urchin	O3:K6
Iwate Prefecture	Sea urchin	O3:K6
B, Iwate Prefecture	Sea urchin	O3:K6
Iwate Prefecture	Sea urchin	O3:K6
China	Sea urchin	O3:K6 and others
China	Sea urchin	O3:K6
City B or C, Hokkaido	Seafood for *sushi*	O3:K6
A, Iwate Prefecture	Squid	O3:K6
Surrounding Saishu Island	Squid	O3:K6
Fukushima Prefecture	Surf clam	O3:K6
Pacific Ocean offshore of Miyagi Prefecture	Tuna	O3:K6
Tottori Prefecture	Turban shell	O3:K6

SOURCE: Anon., 2001c.

Table III–2 shows that mackerel was one of leading causes of *V. parahaemolyticus* outbreaks in Thailand (Atthasampunna, 1974). Mackerel is harvested and consumed in many countries. World catches of horse mackerel were reported to be 6.5 million tonne in 1995, with the EU, led by the Netherlands and Ireland, as the largest exporters. African countries, particularly Egypt, Nigeria and Côte d'Ivoire, are by far the largest markets for horse mackerel of European origin. Japan, with a long tradition of fishing horse mackerel, is another large market that is dependent on both domestic catches and imports to meet domestic demand. Also, the Caribbean region, led by Cuba, imports horse mackerel of European origin. Horse mackerel is an important fish in both international trade and domestic consumption (FAO, 2000).

V. parahaemolyticus is a natural inhabitant of the marine environment, both seawater and sediment, and is transmitted to humans as a result of consumption of contaminated raw or insufficiently cooked seafood (Matte et al., 1994). Horse mackerel could be contaminated at the port during landing by washing with contaminated seawater. In addition, contaminated raw seafood could bring *V. parahaemolyticus* into kitchen environments, where *V. parahaemolyticus* is disseminated through kitchen utensils such as cutting boards and knifes, and by human hands. Once ready-to-eat food is contaminated in the kitchen through utensils and kept in warm conditions (e.g. >30°C) for more than 2 hours, the concentration of the *V. parahaemolyticus* increases drastically, and the consequent risk of foodborne infection also increases.

Table III–2. Seafood involved in 51 *V. parahaemolyticus* cases in Thailand.

Seafood	No. of cases
Shellfish	13
Mackerel	12
Crab	10
Salted mangrove crab	6
Squid	3
Other sea fish	3
Fish products	4

SOURCE: Atthasampunna, 1974.

Seawater temperature is the most important factor controlling environmental levels of *V. parahaemolyticus,* with densities increasing as temperatures rise from 10°C to 30°C (Anon., 2001a). Therefore most human *V. parahaemolyticus* infections occur in the summer (Anon., 2001c).

Sushi

Sashimi

Figure III–1. *Sushi* and *sashimi*.

III–4. Hazard Characterization

III–4.1 Pathogenicity

Strains of *V. parahaemolyticus* induce beta-type haemolysis when grown on a special blood agar, the so-called Kanagawa Phenomenon (KP), and this has been used as an indicator for pathogenic vibrios for a few decades. KP was shown to be caused by thermostable direct haemolysin (TDH) produced extracellularly by *V. parahaemolyticus* (Nishibuchi, Kumagai and Kaper, 1991). However, some KP-negative strains isolated from clinical sources were shown to produce a TDH-related haemolysin (TRH) but not TDH (Honda, Ni and Miwatani, 1988). Recently, strains capable of producing TDH, TRH or both have been considered as pathogenic strains (Nishibuchi and Kaper, 1990). While prevalence of pathogenic strains in seafood or environmental samples has been found to be relatively low, prevalence of these strains in clinical samples is high (Wagatsuma, 1974; Matte et al., 1994; Wong, Ting and Shien, 1992; Wong et al., 1999; Fang, Huang and Chen, 1987; Chowdhury et al., 2001). Available data on the proportion of pathogenic *V. parahaemolyticus* isolated from environmental and food samples are summarized in Table III–3. Recent advances in DNA colony hybridization and PCR (McCarthy et al., 2000; Blackstone et al., 2003) have facilitated detection of pathogenic *V. parahaemolyticus*. While the prevalence of *tdh*+ isolates in total isolates of *V. parahaemolyticus* was similar to that reported in earlier studies (<1%), approximately 50% of oyster samples (10–12 oysters pooled) or individual oysters were *tdh*+ by either PCR or DNA colony hybridization (Kaufman et al., 2003; Nordstrom et al., 2003; Blackstone et al., 2003). Hara-Kudo et al. (2001) indicated that factors making it difficult to isolate virulent strains of *V. parahaemolyticus* from environmental samples such as seafood and coastal water include: (i) various non-*V. parahaemolyticus* bacteria distributed in the coastal environment can grow in enrichment media and on selective agar media commonly used for isolation of *V. parahaemolyticus*; (ii) both virulent and avirulent strains of *V. parahaemolyticus* inhabit the coastal environment; and (iii) the proportion of virulent strains of *V. parahaemolyticus* in the coastal environment is very low. Hara-Kudo et al. (2003) also showed that the incidence of *tdh*+ *V. parahaemolyticus* tended to be high in samples contaminated with relatively high levels of total *V. parahaemolyticus*. Gyobu et al. (2004) reported that the prevalence of pathogenic strains in fish was higher (40%) in the summer (late August), but lower (0%) in the autumn (late September) in Toyama Prefecture in Japan.

Several reports have been published on virulence factors of *V. parahaemolyticus* other than TDH or TRH. For example, a novel siderophore (Yamamoto et al., 1994, 1995) and lateral flagella (McCarter and Wright, 1993) have been reported to be related to the virulence factors. These factors have been well characterized. They may play some auxiliary roles in pathogenesis of *V. parahaemolyticus*, although direct evidence for their implication in pathogenesis has not been obtained.

Table III–3. Prevalence of *tdh*+ and *trh*+ *V. parahaemolyticus* at harvest, retail and post-cooking.

Sample	% of pathogenic Vp	Pathogenicity indicator	Source
Seawater/mud/oyster	0.35% (47/13345)	KP	Wagatsuma. 1974
Seawater/mud/oyster	0% (0/317)	KP	Kiiyukia et al., 1989
Seawater/oyster	4% (2/50)	TDH	DePaola et al., 1990
Shellfish	3.2%	TDH	Anon., 2001c
Oyster	0.26% (9/3429)	TDH	Cook et al., 2002
Mussels (*Perna perna*)	0.51%	KP	Matte et al., 1994
Seafood	1.65% (2/121)	TDH	Wong, Ting and Shien, 1992
Seafood	0% (0/182)	KP	Fang, Huang and Chen, 1987
Imported Seafood	0%	TDH/TRH	Wong et al., 1999
Imported seafood (raw)	1.99% (14/705)	TDH/TRH	Chowdhury et al., 2001
Imported seafood (frozen)	0.84% (5/598)	TDH/TRH	Chowdhury et al., 2001
Seafood	0.68% (2/296)	TDH/TRH	Vuddhakul et al., 2000
Aquatic environment	0.33% (5/1500)	TDH	Islam et al., 2004

A new clone of *V. parahaemolyticus*, serotype O3:K6, has emerged in south Asia and has been the causative pathogen in several outbreaks in Asian countries (Okuda et al., 1997). These O3:K6 strains carry the *tdh* gene, but not the *trh* gene, and show identical DNA fingerprints when examined by an arbitrarily primed-PCR (APPCR) method (Matsumoto et al., 2000).

The differences between this serotype and others were not clearly recognized in terms of their virulence. A number of reports have been published on the characteristics of this pandemic clone, indicating that this clone has some unique properties that contribute to it being a pandemic clone, such as adherence and cytotoxicity (Yeung et al., 2002), low swarming ability, survival at low temperature (5°C, -20°C), and survival under high temperature after heat shock, while increasing production of outer membrane protein of 17 kDa (Kamruzzaman et al., 2006).

A recent study on the full genome sequence of Kanagawa-positive Serotype O3:K6 strain revealed that the genome of the strain has genes for the type III secretion system (TTSS). The TTSS is a central virulence factor of diarrhoea-causing bacteria such as *Shigella*, *Salmonella* and enteropathogenic *Escherichia coli*, and the TTSS is considered as one of pathogenic mechanisms of *V. parahaemolyticus* (Makino et al., 2003).

III–4.2 Dose-response relationship

There are limited numbers of data sources on the infectious dose of *V. parahaemolyticus*. A feeding study of human volunteers showed that even high concentrations, up to 10^{10} cells of KP-negative strains, did not cause any symptoms in 4 healthy volunteers. When KP-positive strains were ingested at three dose levels, 2×10^2, 2×10^5 and 3×10^7, symptomatic cases resulted in 0, 1 and 2 volunteers, respectively, (out of 4 healthy volunteers) (Sanyal and Sen, 1974). The Food and Drug Administration of the United States (USFDA) (Anon., 2001a) developed a dose-response model using data sets from human volunteer studies. This dose-response model was also applied in the hazard characterization of the current risk assessment. The Beta-Poisson dose-response relationship is described as follows:

$$\Pr(ill \mid d) = 1 - \left(1 + \frac{d}{\beta}\right)^{-\alpha}$$

where d = dose of the virulent *V. parahaemolyticus*, and α and β are parameters determined from the data. The food-specific adjustment factor that was introduced for the USFDA risk assessment of raw oysters (FDA, 2005) is not included and the relationship determined from human-feeding tests is used here.

Epidemiological data on the sex and age distribution of the reported cases of *V. parahaemolyticus* infection in Japan showed that the cases were distributed from young children to elderly people, and peaked in the middle-aged group (50–59-year old) without significant differences between men and women (Figure III–2)(Anon., 1998b). A distribution of the affected age group was almost consistent with the distribution of the age group of people consuming raw horse mackerel. The infection was usually considered to cause mild gastro-enteritis 4–30 hours after exposure. Cases of *V. parahaemolyticus* bacteraemia were reported among the patients who were immunosuppressed, especially with leukaemia and cirrhosis (Ng et al., 1999).

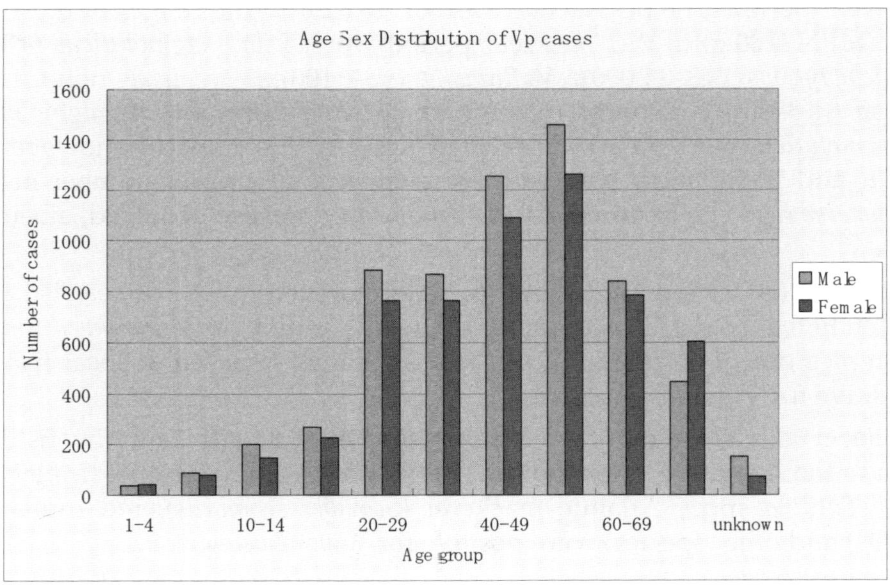

Figure III–2. Age distribution of patients infected with *V. parahaemolyticus* (Anon., 1998b)

III–5. Exposure Assessment

III–5.1 Prevalence in foods

Although limited information is available on foods associated with *V. parahaemolyticus* outbreaks in countries other than Japan, there are several reports on the high prevalence of the organism in a variety of seafood.

Wong, Chen and Yu (1995) reported that *V. parahaemolyticus* was detected from 1 sample out of 34 (2.9%) frozen shrimp dumplings, 7 samples out of 40 (17.5%) frozen fish dumplings and 10 samples out of 40 (25.0%) frozen peeled shrimps. Wong et al. (1999) also reported that *V. parahaemolyticus* was recovered from 315 (45.9%) fresh raw imported seafood samples. The incidence of *V. parahaemolyticus* in products from Hong Kong and Thailand was markedly higher than the incidence in products from Indonesia and Viet Nam. The percentage prevalence in shrimp, crab, snail, lobster, sand crab, fish and crawfish were 75.8, 73.3, 44.3, 44.1, 32.5, 29.3 and 21.1%, respectively (Wong et al., 1999).

Matte et al. (1994) examined mussels (*Perna perna*) harvested on the coast of Ubatuba at three different stations in the State of Sao Paulo, Brazil, for *Vibrio* spp. over a 1-year period. The prevalence was 66.7 to 92.0%. The ranges of most probable number (MPN/100 g) in this one-year period were: *Vibrio alginolyticus* (<3–24 000), *V. parahaemolyticus* (<3–24 000), *V. fluvialis* (<3–1100), *V. cholerae* non-O1 (<3–23), *V. furnissii* (< 3–30), *V. mimicus* (<3–9) and *V. vulnificus* (<3–3). *V. alginolyticus* and *V. parahaemolyticus* were the most prevalent species. Other species, however, such as *V. fluvialis* and *V. vulnificus*, were also present in a significant number of samples, and *V. cholerae* non-O1, *V. furnissii* and *V. mimicus* were also observed, albeit at lower levels.

Jaksic et al. (2002) reported that 1 sample out of 10 (10%) of sea fish intestines sampled in hotels on the Croatian sea coast were positive for *V. parahaemolyticus*, and three samples out of 50 (6.0%) of sea fish intestines sampled at local fish markets were positive for *V. parahaemolyticus*.

Baffone et al. (2000) reported 3 samples out of 114 (2.6%) raw fish samples common to the Adriatic Sea (anchovies (*Engraulis* spp.), grey mullet (*Mugil cephalus*), sardines (*Sardina* spp.), Atlantic mackerel (*Scomber scombrus*), red mullet (*Mullus surmuletus*) and other species) were positive for *V. parahaemolyticus*.

Chan et al. (1989) reported summer prevalence of *V. parahaemolyticus* and other halophilic vibrios in seafood from Hong Kong markets. Halophilic vibrios were isolated from all seven types of seafood examined, and comprised 9.1%, 8% and 6.1% of contaminating aerobic heterotrophic bacteria from mussels, clams and oysters, respectively.

El-Sahn, El-Banna and El-Tabey Shehata (1982) examined samples of seawater and sediment invertebrates from around Alexandria, Egypt. Average counts (per 100 mL or 100 g) of *V. parahaemolyticus* were as follows: seawater, 36; sea urchins, 349; sediment, 436; wedge shells, 534; and clams, 1872. Samples collected in the summer months contained higher levels of *V. parahaemolyticus* than winter samples.

Chowdhury et al. (2001) reported the prevalence of *V. parahaemolyticus* in imported frozen seafood sampled at Osaka port and imported fresh seafood sampled at Kansai international airport during 1998–2000. Out of the 335 frozen samples examined, 593 strains of *V. parahaemolyticus* were isolated from 65 samples (19.4%) (on average, 8.9 strains of *V. parahaemolyticus* were isolated from one sample), and out of the 949 fresh samples, 705 strains of *V. parahaemolyticus* were isolated from 234 samples (25%) (on average, 3 0 strains of *V. parahaemolyticus* were isolated from one sample). They also reported the differences in prevalence among seafood species and country of origin. Tuna had the highest prevalence among several different species of fresh seafood, and shrimp had the highest prevalence among frozen seafood, while Spanish mackerel had a lower prevalence. They also reported that 2 strains (0.15%) contained the *tdh* gene and 17 strains (1.3%) contained the *trh* gene out of 1298 *V. parahaemolyticus* strains isolated.

III–5.2 Factors influencing the concentration of Vibrio *parahaemolyticus in seawater, environment and food*

The most important factor influencing the density of *V. parahaemolyticus* in seawater is temperature, with 14–20°C considered a threshold range for the organism to leave the viable but non culturable (VBNC) state and begin proliferation (Kaneko and Colwell, 1975). In addition, Ogawa et al. (1989) also reported on the ecology of *V. parahaemolyticus* in Hiroshima Bay and created a logistic regression model to predict numbers of *V. parahaemolyticus* in raw oyster from seawater temperature and seawater salinity (Appendix III–1 contains a summary of the Ogawa report, as it was written in Japanese).

Kumazawa et al.(1999) reported that thick accumulation of muddy sediments on the riverbed and stagnation of brackish water at low tide seem to be essential for *V. parahaemolyticus* to survive in neritid gastropods, including *Clithon retropictus.* Ogasawara (2000) observed that the area where the water flow was very slow and turbid showed a higher density of *V. parahaemolyticus* and longer periods of contamination.

Sarkar et al. (1985) reported that the incidence and counts of *V. parahaemolyticus* were consistently higher in association with plankton than with water and sediment samples. The detection rate of *V. parahaemolyticus* among external surface, gills and faeces from freshly caught freshwater fishes examined from the positive samples was highest in the faecal samples (82.1%), whereas the frequency of isolation from the external surface of the freshwater fishes was lowest (25%).

Watkins and Cabelli (1985) reported that densities of *V. parahaemolyticus* were greatest in the near-surface waters of contaminated areas, and decreased sharply with both the distance from the sources of faecal pollution and the depth of the water column. A positive association with the amount of particulate matter in the water, and specifically with its zooplankton content, was also reported.

A variety of fish species are harvested commercially by several methods of fishing. Coastal fish are caught by trawl and fixed net. Tuna and large deep-sea fish are caught by long line. Some species (Greater amberjack, Red seabream) are farmed in coastal waters. Vessels for long-line fishing are relatively large and usually have refrigeration systems. Also, tuna and other species of fish caught by long lining are usually caught well away from the coast, reducing the opportunity for contamination with *V. parahaemolyticus*.

Fishing vessels for coastal fishing are relatively small. Fishing trips are usually short (several hours) and some of these vessels have no refrigeration. The possibility of contamination and growth of *V. parahaemolyticus* during fishing for coastal fish is higher than for deep-sea fishing. Fish from aquaculture are also more likely to be contaminated with *V. parahaemolyticus*. There are few data on the risk of fish becoming contaminated with *V. parahaemolyticus* on board fishing vessels. In an International Commission on Microbiological Specifications of Food (ICMSF) report (Anon., 1988), gilling and evisceration were found to be done immediately after catching. This was undertaken by hand using a sharp knife to release the intestines, including the digestive tract, which represents a reservoir for potential spoilage bacteria. However, fish processing is not commonly undertaken on the fishing vessels. The risk of contamination associated with this step is not fully understood.

Sakazaki and Nakanishi (1975) reported the prevalence of *V. parahaemolyticus* in several kinds of fish and shellfish, and concluded that the frequency of contamination of shellfish was greater than that of fish, and among fish types, those with scales were more frequently contaminated than those without scales.

Yamazaki et al. (1996) reported the prevalence of *V. parahaemolyticus* in various kinds of fish and concluded that inshore fish are more likely to be contaminated with *V. parahaemolyticus* than are fish caught in deeper waters. Cultured fish, such as greater amberjack and red sea bream, were also contaminated with *V. parahaemolyticus*.

Shiozawa et al. (1998) undertook a survey on the density of *V. parahaemolyticus* in contaminated house mackerel, clam and round clam on a monthly basis. Clam and round clam were highly contaminated (about 2 $\log_{10}$ higher than horse mackerel and seawater). The same authors studied the relationship between the levels of *V. parahaemolyticus* in seawater and those in fish (horse mackerel). Horse mackerel had lower levels of *V. parahaemolyticus* than raw oyster.

III–5.3 Concentration of Vibrio parahaemolyticus in horse mackerel

Quantitative data for *V. parahaemolyticus* in fish in the English-language literature is very limited. Chan et al. (1989) reported that the concentration of *V. parahaemolyticus* in portions of gills, gut, skin and muscle of rabbit fish was 82 cfu/g, and for the same portions of grouper fish it was 88 cfu/g.

For the development of a quantitative risk assessment model, Japanese published and unpublished data were used on the prevalence and concentration of *V. parahaemolyticus* in horse mackerel at different steps in the food chain.

III–5.4 Production to consumption pathway

Japanese horse mackerel (*Trachurus japonicus*) is caught in the waters around Japan, in the East China Sea, and in various areas around Africa. In Japan, 30% of horse mackerel is consumed raw as *sushi* or *sashimi*. In 2000, 282 000 t of horse mackerel were landed from the sea around Japan (Anon., 2001b). One harvesting trip usually takes one to two days. Just after catching, the fish are transferred to a transportation vessel, which has a holding tank with iced seawater. The transportation vessel commutes between the harvesting areas and a landing place, and fish are transported as soon as a fish hold is full. Upon arrival at the landing place, the fish are often washed with seawater during the landing operation and then sorted by species and size. They are then kept in boxes (5–7 kg) filled with ice, for auction in the onshore market.

Landing fish from the fishing boat

Transportation to the onshore market after landing

The onshore market

The market near to a large city

A retailer

Figure III–3. From harvest to retail.

Upon arrival at the landing place, some fish are transported directly from the transport vessel to a large container (200 kg) or truck, and then auctioned. In these cases, fish are transported to processing establishments near the port. The fish purchased by wholesalers are transported to the markets of nearby large cities or directly to retailers. During this on-land transportation, fish are kept in boxes with ice, and transported in refrigerated trucks. This transportation takes a maximum of 2 days. Auctions near large cities begin early in the morning (e.g. between 04:00 and 06:00), and then intermediaries, who purchase the fish, transfer the fish to retailers before 10:00. The stages are illustrated in Figures III–3 and III–4.

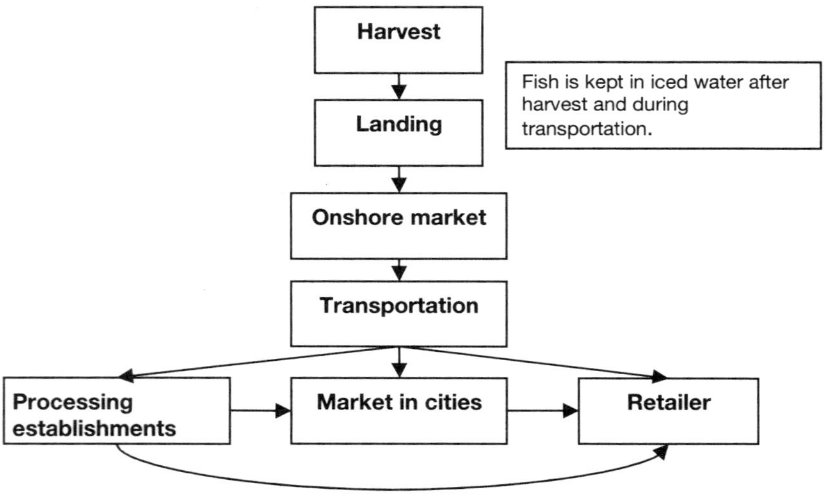

Figure III–4. Flow diagram from harvest to retailer.

III–5.5 Data for pre-harvest and harvest

There were data available on the concentration of *V. parahaemolyticus* in the gills of horse mackerel, which were obtained during 2001–2002 at selected public health institutes of Miyagi, Tokyo, Kanagawa, Shizuoka, Shimane and Kumamoto Prefectures, Japan (Ken Osaka, pers. comm.). These data are shown in Figure III–5, together with estimates of the surface sea water temperature at harvest. The data set by Ohno et al. (1993) was the only available data set that reported concentration of *V. parahaemolyticus* in each part of the horse mackerel, i.e. surface, gills and intestine, although only the mean numbers were reported. In addition, this data set was generated during summer and autumn, when seawater temperature is high and consequently the concentration of *V. parahaemolyticus* in seawater and fish is also high. Therefore, we could consider that this data set was generated during months when foodborne illnesses caused by *V. parahaemolyticus* occur frequently, and therefore it could be considered as representative of the worst-case scenario.

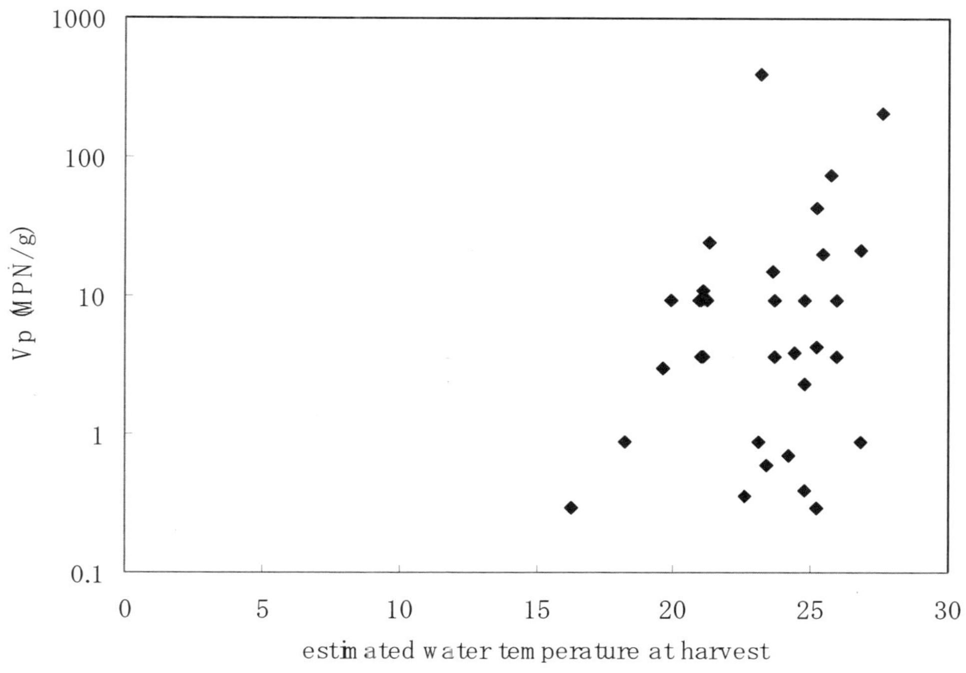

Figure III–5. Relationship between estimated water temperature (°C) and the concentration (MPN/g) of *V. parahaemolyticus* in gills of horse mackerel.

III–5.6 Handling at port and transportation

III–5.6.1 Contamination caused by water used during landing and selling at port market

During landing at the port, there are several potential opportunities for fish to become contaminated by the water used. Water is used during landing and sale:

- during selling fish at market;
- for storage of live fish; and
- for transportation of fish.

Yamai (2001) examined water samples that were used during selling (washing fish), transporting and preserving live fish. The study showed that water used during the process of landing, selling and transportation was highly contaminated. Some water was disinfected through powdered activated carbon (PAC) coagulation and filtration, chlorination and UV irradiation. As a result of the study it was found that 73% of samples of water used without any treatment or processing were contaminated with *V. parahaemolyticus*, 38% of samples were contaminated at a level of 10^3–10^5 MPN/100 mL, 71% of water treated with PAC coagulation and filtration was contaminated at a relatively low level, and 92% of disinfected water was found to be uncontaminated.

There were two data sets available for the port and market stages post-harvest. One was a survey of seawater actually undertaken in local landing places and port markets. The data included information on the concentration of *V. parahaemolyticus* in water used. The water used was either natural seawater pumped up from nearby coastal water, disinfected seawater (using processes such as PAC coagulation), or seawater irradiated with ultraviolet light. An additional data set was generated by Kumagai et al. (2003) for this risk assessment, in collaboration with WHO. They carried out several experimental tests in the laboratory in order to simulate actual fish washing procedures at landing places and markets: this involved immersing fish in natural or artificially composed seawater with and without specific levels of *V. parahaemolyticus*. The data showed that effectiveness of washing with clean water was minimal for reducing the number of contaminated bacteria in gills and intestines, while bacterial levels on the surface of finfish were considerably reduced (Table III–4).

Table III–4. Effectiveness of immersing the horse mackerel in clean water for reducing the concentration of *V. parahaemolyticus* on/in skin surface, gill and intestines.

	Mean	Std. dev.
Skin surface (cfu/cm^2)		
A: concentration of *V. parahaemolyticus* on skin surface immediately after immersing fish into seawater with 40 000 cfu/mL for 1 minute	78.42	23.56
B: concentration of *V. parahaemolyticus* on skin surface immediately after immersing fish in clean water for 3 hours	12.36	9.20
B/A reduction ratio	0.158	0.126
Gill (cfu/g)		
A: concentration of *V. parahaemolyticus* in gills immediately after immersing fish into seawater with 40 000 cfu/mL for 1 minute	4.09	1.27
B: concentration of *V. parahaemolyticus* in gills immediately after immersing fish in clean water for 3 hours	5.52	4.26
B/A reduction ratio	1.350	1.122
Intestines (cfu/g)		
A: concentration of *V. parahaemolyticus* in intestines immediately after immersing fish into seawater with 40 000 cfu/mL for 1 minute	2.70	3.09
B: concentration of *V. parahaemolyticus* in intestines immediately after immersing fish in clean water for 3 hours	2.36	5.11
B/A reduction ratio	0.873	2.140

III–5.6.2 Storage and mitigating processes

Temperature

There are several reports on temperature and the growth of *V. parahaemolyticus*. Survival of *V. parahaemolyticus* was determined in oyster meat homogenates at various temperatures. (4°C, 0°C, -18°C and -24°C) and bacterial levels (10^2, 10^4, 10^5 and 10^7/mL). In all cases, the numbers of *V. parahaemolyticus* were a logarithmic function of log time. This study indicated that *V. parahaemolyticus* can be inactivated at low temperatures. The time taken for total inactivation depends on the initial number of micro-organisms and the incubation temperature (Muntada-Garriga et al., 1995).

Oliver (1981) compared survival of *Vibrio vulnificus* and *V. parahaemolyticus* in oyster homogenates held at 4°C. Their results indicated a rapid decrease in viability, not attributable to either cold shock or the oyster homogenate alone but to a combination of the two for *V. vulnificus,* but such a decline was not observed with *V. parahaemolyticus*.

Miles et al. (1997) reported the growth rates of four strains of *V. parahaemolyticus* that were measured and compared in a model broth system. The results for the fastest growing strain, based on 77 combinations of temperature and water activity (a_w) using NaCl as humectant, were summarized in the form of a predictive mathematical model. The minimum temperature observed for growth was 8.3°C, while the maximum temperature for growth was observed at 45.3°C, with the optimum occurring between 37 and 39°C.

Hiro et al. (1996) measured the growth of *V. parahaemolyticus* in round clam and turban shellfish at 10°C and 25°C. While *V. parahaemolyticus* in unshucked shellfish did not grow appreciably, *V. parahaemolyticus* on the meat of round clams increased by one Log in 6 hours at 25°C.

Watanabe (1994) reported similar data for *V. parahaemolyticus* on horse mackerel, with an increase from 10 to 10^3 times in 8 hours at 25°C, but there was no significant growth in 4 hours.

Iawashita (1991) had investigated the situation of *V. parahaemolyticus* contamination at each step during transport and processing in 1988–1990 in Yamanashi Prefecture, Japan. In the study, many wholesalers sold fish and shellfish for raw consumption within 3 days after they stocked them. The refrigeration temperatures in these facilities were investigated and it was found that all of them were below 5°C except in the case of one large retailer. They also measured the level of *V. parahaemolyticus* in the fish at each of the steps from transport from the local fisheries market to retailers over a number of years. They analysed the levels of *V. parahaemolyticus* in round clams of the same lot at each stage during distribution, and concluded that if each step of transportation were kept at less than 10°C, the increase in *V. parahaemolyticus* would be negligible.

pH and other factors

V. parahaemolyticus has been shown to grow at pH 5–11 and at NaCl concentrations of 1–7% (Twedt, Spaulding and Hall, 1969). Beuchat (1973) reported that some strains of *V. parahaemolyticus* could grow at pH 4.8.

Other possible factors inhibitory to the growth of *V. parahaemolyticus* have been reported, such as glycerin (Chun et al., 1972), high pressure (Baross, Hanus and Morita, 1975), UV irradiation (Hackney, Ray and Speck, 1988), basil and sage essential oils (Koga, Hirota and Takumi, 1999) and mild heating (Beuchat and Worthington, 1976). However, there is little potential for control measures based on these to be implemented during the process of harvesting and consuming raw fish.

III–5.7 Preparation and consumption

III–5.7.1 Setting of outbreaks

A ten-year summary of outbreaks due to *V. parahaemolyticus* in Japan (Anon., 2000a) is shown in Figure III–6 and shows that outbreaks due to *V. parahaemolyticus* occurred in restaurants (48%), Japanese traditional hotels (18%), catering and lunch boxes (12%), home (12%) and others (10%). Horse mackerel is one of the most popular finfish throughout the year, which is often purchased at retail stores as a whole fish, then prepared and consumed in households and restaurants.

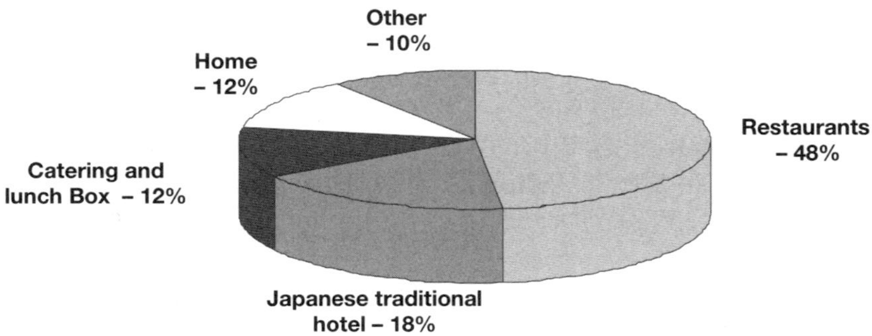

Figure III–6. Settings of outbreaks due to *V. parahaemolyticus*.

III–5.7.2 Preparation procedure – washing finfish with tap water and methods of preparation of *sashimi*

Watanabe (1994) studied different procedures for the preparation of horse mackerel and tested the effectiveness of washing the fish with tap water as a means of reducing the bacterial load. The results of an experimental study show that washing fish with tap water before preparing could reduce the level of *V. parahaemolyticus* on the fish surface, but that it did not reduce the levels in the gills and intestines (Figure III–7).

The study also showed that washing with tap water during the process of making *sashimi* was important. Two preparation procedures were compared: the first one involved washing the eviscerated cavity and in the second scenario the cavity was not washed. When compared, it was found that the *sashimi* prepared without washing of the eviscerated cavity was highly contaminated with *V. parahaemolyticus* (Figure III–8). In the case of horse mackerel *sashimi*, only pieces of fish fillet without bones and intestines are usually eaten. The fish fillet (muscle) is not considered to be contaminated before preparation started, but is contaminated during the preparation process from contaminated gills and intestines via the cutting board and knife.

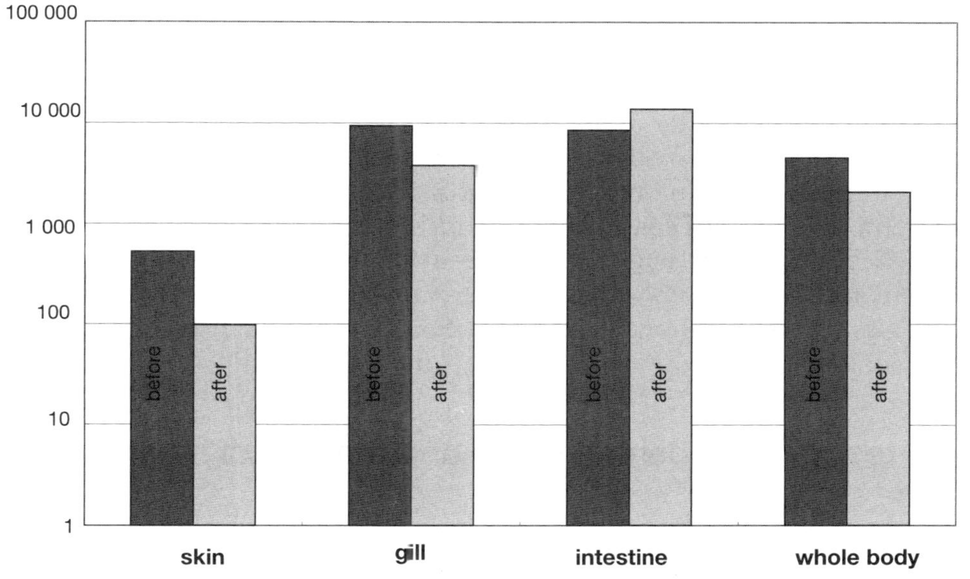

Figure III–7. Effectiveness of washing the finfish body with tap water for reducing the concentration of *V. parahaemolyticus* on/in skin surface, gills, intestines and whole body. Units are MPN/g for gill and intestine; and MPN/cm² for skin surface and whole body.

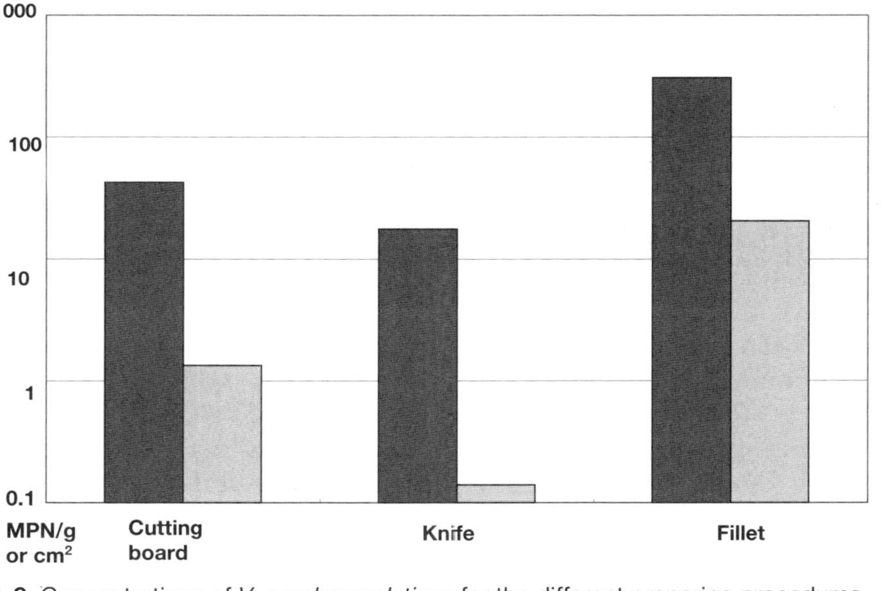

Figure III–8. Concentrations of *V. parahaemolyticus* for the different preparing procedures.
KEY: A (dark grey): preparing without washing eviscerated cavity with tap water;
B (light grey): preparing with washing eviscerated cavity with tap water.
SOURCE: Watanaba, 1994.

III–5.7.3 Time between preparation and consumption

There were no available data on time between preparation and consumption. Most people like to eat fresh food, and especially fresh raw fish, so the fish is usually consumed within one hour after preparation at home. It may sometimes happen that in large restaurants, hotels and catering services the *sashimi* is prepared several hours prior to consumption, but in most such cases the raw fish (*sashimi*) is usually refrigerated until the point of consumption. Thus, under such conditions, i.e. *sashimi* stored in the refrigerator, the length of time between preparation and consumption does not seem to be critical in terms of bacterial growth. Many people purchase ready-to-eat *sashimi* combinations at supermarkets, where raw fish and *sashimi* are required to be stored at below 10°C (with below 4°C strongly recommended) in Japan.

III–5.7.4 Frequency of consumption and amount of raw fish consumed

There were limited data on consumption of raw fish such as *sushi* and *sashimi*. An official report on the household budget reported that one household in Japan purchased between 45 and 50 kg of fresh fish and shellfish and 2.2 to 3 kg of a combination of different species of *sashimi* (very thin bite-size slices of fresh raw fish) per year (Anon., 2000b).

Another report on nutrition revealed that the frequency of eating *sushi* was 5.9% per one meal out of home, and the frequency of eating out was 16.8% (breakfast, 2.7%; lunch, 40.5%; dinner, 7.3%) (Anon., 1998a). One large-scale nutrition survey involving 15 000 people demonstrated the consumption of raw horse mackerel in a single day of November (Anon., 1995). The results showed that 59 individuals out of 14 240 ate raw horse mackerel (414 per 100 000) with the serving size ranging from 2.5 to 250 g and an average serving size of 73 g, as shown in Figure III–9.

By using the values given in Section 5.4 above, i.e. (i) the annual harvest weight of horse mackerel from surrounding water of Japan: 282 000 t; (ii) 30% of horse mackerel is consumed raw, and the value from the previous section indicating that (iii) the average consumption weight per serving is 73 g; and considering the total Japanese population is 1.2×10^8, the number of consumers of raw horse mackerels per 100 000 population was calculated as

$$(2.82\times10^{11}\text{ g} \times 0.30)/(73\text{ g} \times 365\text{ days} \times 1.2\times10^8) = 2.65\times10^{-2} = 2650 \text{ per } 100\,000$$

which is 6.4 times greater than the estimate above (414 per 100 000) based on a nutrition survey (Anon., 1995). It was considered that the difference was due to the difference in the frequency of the raw consumption of horse mackerel in certain months; the frequency of consumption of raw horse mackerel is generally considered to be smaller in November than other months.

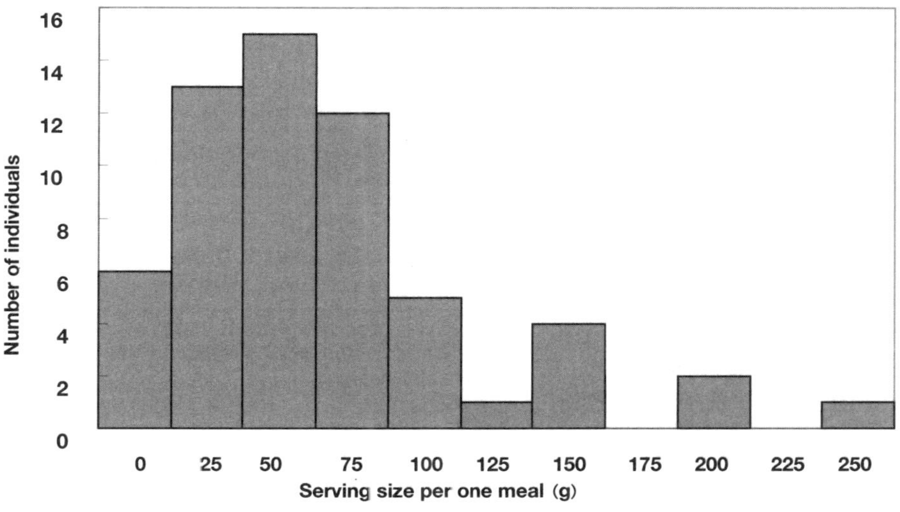

Figure III-9. Consumption size of raw horse mackerel per single day. Based on data from Anon., 1995.

III–5.8 Risk assessment model structures

III–5.8.1 Scenario

In this risk assessment, two scenarios were considered at each of the landing places (washing with clean water or not), the transport stage (using clean water or contaminated water) and the preparation stage (washing the eviscerated cavity or not) (Figure III–10). In total, eight different scenarios were considered and each scenario was simulated with Monte-Carlo methodology using @RISK®™ software.

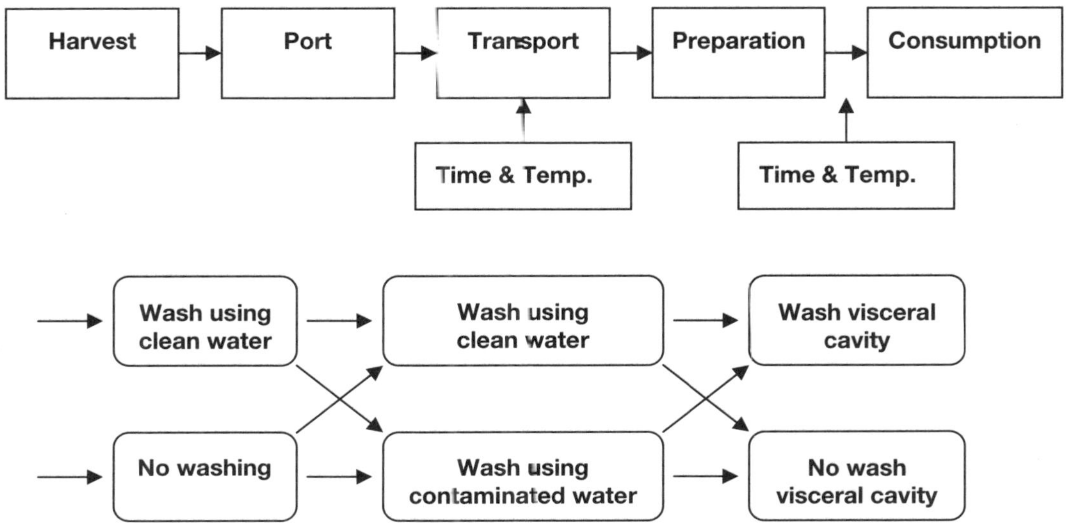

Figure III-10. Proposed scenarios at each stage and the eight scenarios modelled in the finfish (horse mackerel) risk assessment.

III–5.8.2 Harvest stage

The concentration of *V. parahaemolyticus* in gills at harvest was simulated by sampling the data by Nakajima (pers. comm., 2004) and Prefecture public health institutes, as shown in Table III–5. Only the data obtained from June to October—when most of the foodborne illnesses caused by *V. parahaemolyticus* occur in Japan—for fish sampled at onshore fish markets were used as the data at harvest.

Table III–5. Concentration of *V. parahaemolyticus* by part of horse mackerel (N=25).

Part	Mean*	log_{10}(mean)	log_{10}(standard deviation)
Surface	5.1 MPN/cm^2	0.708	
Gills	660 MPN/g	2.820	0.767**
Intestines	1300 MPN/g	3.114	

NOTES: *Only means were reported by Ohno et al. (1993). **Calculated from the data by Nakajima (2004) and Prefecture public health institutes.

Since the fish sample numbers per month varied among prefectures, the sampling frequency in the simulation was defined in such a way as to give equal weight to data for each location and month. If there are multiple data for a specific month, then numbers of *V. parahaemolyticus* per gram are multiplied by 1/number of data available for the specific month, e.g. 3 data = 1/3 = 0.33; 4 data = 1/4 = 0.25.

For examples, in June in Kanagawa Prefecture, there are 3 data values (29, 9 and 0 Vp/g), therefore the representative concentration of *V. parahaemolyticus* in June in Kanagawa is $29\times(1/3) + 9\times(1/3) + 0\times(1/3) = 12.67$.

It was then assumed that the levels of *V. parahaemolyticus* on the surface of the fish and in the intestines at harvest were proportional to that from the gill, with the ratio calculated from the data reported by Ohno et al. (1993) (Table III–6). This report was the only available data set for all the fish parts of interest: gills, intestine and skin surface, and the data were collected in all seasons, including summer.

Table III-6. Concentration of *Vibrio parahaemolyticus* in gills.

Harvest place	Month	Vp/g	Data weight for simulation
Kanagawa Prefecture	June	29	0.333
		9	0.333
		0	0.333
	July	1.1	0.25
		0	0.25
		0	0.25
		0	0.25
	August	7	0.25
		0.4	0.25
		0.3	0.25
		0.3	0.25
	September	0.4	0.333
		0	0.333
		0	0.333
	October	0	1
Shizuoka Prefecture (cultured)	June	0	1
	July	3000	1
	August	5500	1
	September	940	1
	October	16	1
Shizuoka Prefecture (natural)	June	0.36	1
	July	2.3	1
	August	0	1
	September	20	1
	October	0	1
Kumamoto Prefecture	July	390	1
	August	0	1
	September	210	1
Shimane Prefecture	August	0.9	1
	September	4.3	1

NOTE: Data reported by Prefecture public health laboratories, except for that from Kanagawa Prefecture (The number of *Vibrio parahaemolyticus* and prevalence of TDH and TRH genes from horse mackerels harvested in Kanagawa Prefecture, T. Nakajima, pers. comm., 2004).

III–5.8.3 Post-harvest stage

Port landing

Two scenarios were simulated for this stage: washing fish with vibrio-free water (scenario 1) or no washing (scenario 2). The effect of washing fish with potable water was modelled based on the data of the concentration of *V. parahaemolyticus* before and after the landing stage as reported by Ohno et al. (1993). The value of the ratio was derived from concentration after washing divided by concentration before washing, and calculated for surface and gills separately (Table III–7). The mean and standard deviation of their ratios were calculated from this data, and a normal distribution was assumed for the ratio. It was assumed that washing was not effective against *V. parahaemolyticus* present in the intestines.

Table III–7. Effect of washing of fish with potable water: ratio of concentration of *Vibrio parahaemolyticus* after and before washing with potable water.

Part of the fish	Mean	Std dev
Surface	0.038	0.024
Gills	0.401	0.252

NOTE: Ratio is mean concentration after washing divided by mean concentration before wahing.

Storage and transportation from port market to wholesaler and to retailer

Two scenarios were also simulated for this stage: transport of fish in contaminated seawater (Scenario 1) or in clean, vibrio-free seawater (Scenario 2).

For Scenario 1, experimental data (Table III–4) were used to estimate the level of *V. parahaemolyticus* in each part of the horse mackerel. The cross-contamination rate from contaminated seawater during washing was modelled based on the data by Kumagai et al. (2003) (Table III–8), and a normal distribution was assumed. It was assumed that the number of *V. parahaemolyticus* in fish should be proportional to the concentration of *V. parahaemolyticus* in seawater, which, in turn, was assumed to follow a RiskPert distribution (Minimum 1, Maximum 6, Most likely 4) when converted to normal logarithm per 100 mL based on the data shown in Table III–5.

Table III–8. Transfer rate from seawater to horse mackerel.

	Mean	Std dev
$T_{surface}$	0.001960	0.000295
T_{gills}	0.000102	0.000016

NOTES: $T_{surface}$ = Vp on surface (no. per cm^2)/Vp in seawater (no. per mL). Mean concentration on skin surface described in Table III–4 (78.42 cfu/cm^2) was divided by inoculated concentration (40 000 cfu/mL) in seawater.
T_{gills} = Vp in gills (no. per gram)/Vp in seawater (no. per mL). Mean concentration on gills described in Table III–4 (4.09 cfu/g) was divided by inoculated concentration (40 000 cfu/mL) in seawater.

In the storage and transportation stage, available data showed that the air temperature during each stage of transportation was <10°C. A scenario in which the mean temperature was 6°C (range 3–9°C) and mean elapsed time was 36 hours with a range of 6 to 60 hours was used. The FDA-VPRA (FDA, 2005) growth model was applied. The growth of *V. parahaemolyticus* during the post-harvest stage was calculated to be small (approximately 10% growth).

III–5.8.4 Preparation

In order to calculate the total number of *V. parahaemolyticus* in a fillet after preparation, data reported by Ohno et al. (1993) were used, which described the transfer rate of *V. parahaemolyticus* from entire fish body to fillet.

It was assumed that *V. parahaemolyticus* is located mainly on the surface, in gills and intestines, and the total number of *V. parahaemolyticus* in each part was calculated by multiplying the mean concentration by the area (surface; cm^2) or the mean concentration by the weight of gills or intestines (gram), then the sum of them was divided by the body weight to obtain concentration in the entire body. In this step, since the distributions of body weight, gill weight, intestine weight and skin surface area were not known, the fixed values that were reported by Ohno et al. (1993) as shown in Table III–9 was used. Thus the concentration of *V. parahaemolyticus* in the entire fish before preparation was calculated as follows:

Table III–9. Weight and surface area of horse mackerel used in the model.

Body weight	Gill weight	Intestine weight	Skin surface area
80 g	0.7 g	5.6 g	96 cm^2

SOURCE: Based on data from Ohno et al. (1993).

[(mean number on surface (5.1 MPN/cm^2) × skin surface area (96 cm^2)) + (mean number in gills (660 MPN/g) × gill weight (0.7 g)) + (mean number in intestines (1300 MPN/g) × intestinal weight (5.6 g))]/total body weight (80 g).

The effect of washing the eviscerated cavity during preparation was simulated based on experimental data (Table III–8); the concentration of *V. parahaemolyticus* on the fillet after preparation was calculated from the whole body concentration by using the mean and standard deviation of the transfer rate shown in Table III–10.

Table III–10. Transfer rate of *V. parahaemolyticus* from whole fish body to fillet during preparation (ratio of concentration).

Processing	Mean	Std dev.
Without washing	0.2268	0.0930
With washing	0.0132	0.0046

Source: Based on data from Ohno et al. (1993).

III–5.8.5 Consumption module

In most cases of raw horse mackerel consumption, prepared raw fish fillets are consumed shortly after preparation, but in rare cases they are left on tables at room temperature for a few hours before consumption. In developing this module it was therefore assumed that the raw fish was kept at 15 to 32°C (most probably 22°C) for 0.25 to 4 hours (most probably 1 hour) before consumption, and PERT distributions were applied to both input values (see Table III–11). The same growth model was used here as was used for the transportation and storage stage.

Consumption data from the national survey of Japan in 1995 were used to estimate the frequency of consumption and the serving size. It was assumed that the frequency of consumption and the amount of raw horse mackerel consumed (Figure III–9) were stable throughout the year. The amount of raw horse mackerel eaten in a day was simulated by sampling the survey data, and the mean frequency of eating raw horse mackerel was used in conversion to annual risk.

Table III–11. Distributions used in *V. parahaemolyticus* in horse mackerel model modules.

Inputs	Distribution
Harvest	
Log (Vp) surface density (/cm^2)	= Log (Vp gill) – (2.820-0.708), max 3.477
Log (Vp) gill surface density (/cm)	= RiskDiscrete (Table III–5 data)
Log (Vp) intestine density (/cm^3)	= Log (Vp gill) – (2.820-3.114), max 8
Post harvest (landing and washing at port and port market)	
Washing reduction ratio (surface)	= RiskNormal (0.038, 0.024)
Washing reduction ratio (gill)	= RiskNormal (0.401, 0.252)
Log (Vp) in contaminated water	= RiskPERT (1, 4, 6)
Vp transfer from water to fish	
Surface	= RiskNormal (0.00196, 0.000295)
Gill	= RiskNormal (0.000102, 0.000016)
Virulent Vp fraction	= RiskDuniform (Table III–3 data)
Transport & storage	
Total elapsed time (hour)	= RiskPERT(6, 36, 60)
Storage temperature (°C)	= RiskPERT(3, 6, 9)
Storage in clean water reduction ratio (surface)	= RiskNormal (0.158, 0.063)
Preparation – Contamination from whole body to fillet	
Without washing	= RiskNormal (0.2268, 0.0930)
Washing with clean water	= RiskNormal (0.0132, 0.0046)
Consumption	
Elapsed time (hour)	= RiskPERT (0.25, 1, 4)
Storage temperature (°C)	= RiskPERT (15, 22, 32)
Raw fillet weight/serving (g)	= RiskDuniform (Figure III–9 data)

III–5.8.6 Factors not included in the model

The factors listed in Table III–12 should be considered during the development of an exposure assessment model; however, due to the lack of available data and to avoid unnecessary complexity in the model in addressing the risk management question that was being addressed, they were not included in this model.

Table III–12. Factors not included (modelled) in the current model.

Stage	**Factors**
Harvest	Water temperature and salinity Tide Plankton Species of fish Methods of fishery Where the fish are captured Coastal, aquacultured Time and temperature combination from fish capture to port on fishing vessels
Post-harvest (landing place, port market, transport from port market to wholesaler, storage at wholesaler, transport from wholesaler to retailer, storage at retailer)	Time and temperature combination of each different step Time and temperature combination from retailer to home (usually ambient temperature)
Preparation	Cross-contamination during heading and evisceration, from fish to hand, hand to other food, fish to cutting board, then cutting board to other fish. Ratio between home preparation vs commercial preparation in retailers, hotels and restaurants Home storage from retail purchase to consumption.
Consumption	Seasonal and regional difference in raw horse mackerel consumption. Consumption of more than two fish. Fish body size distribution. Difference in preparation process (*tataki* and *sashimi*). Raw fish consumption with vinegar.

III–6. Risk Characterization

Risk characterization is the integration of the exposure assessment and the hazard characterization or dose-response assessment. In summary, the total number of *V. parahaemolyticus* per serving and the number of pathogenic *V. parahaemolyticus* per serving were simulated, and then outcomes were input into the dose-response model. This section describes the probability of illness caused by consumption of raw horse mackerel contaminated with pathogenic *V. parahaemolyticus*. The following tables (Tables III–13 and III–14) present the numbers of illnesses predicted, based on the model assumptions described in the previous sections.

Table III–13. Results of Monte Carlo simulation in each stage of selected scenarios: *V. parahaemolyticus* densities and concentrations.

Scenario			Vp levels		
	Transport	**Harvest**	**Surface**	**Gills**	**Intestines**
			Vp/cm^2	Vp/g	Vp/g
Harvest			3.90	505	995
Post-harvest	No washing		3.90	505	
	Washing		0.15	204	
After transport	Clean water	No washing	0.69	540	1064
		Washing	0.03	219	1064
	Contaminated water	No washing	5.01	540	1064
		Washing	1.00	219	1064

Table III–14. Results of Monte Carlo simulation for selected scenarios: *V. parahaemolyticus* numbers and probability of becoming ill per raw horse mackerel serving.

Scenarios			Vp in whole body	Fillet Vp/g	Virulent Vp/serving	Probability of illness per serving
Harvest	**Transport**	**Preparation**				
No washing	Contaminated water	No washing	6841	19.4	27.1	1.61E-05
		Washing		1.13	1.59	9.32E-07
	Clean water	No washing	6425	18.2	25.5	1.51E-05
		Washing		1.06	1.50	8.80E-07
Washing	Contaminated water	No washing	6225	17.7	24.7	1.47E-05
		Washing		1.03	1.45	8.48E-07
	Clean water	No washing	6134	17.4	24.3	1.44E-05
		Washing		1.02	1.43	8.40E-07

Sensitivity analysis (Figure III–11) indicates that uncertainties in the initial concentration of *V. parahaemolyticus* in fish, the proportion of the virulent strains in total *V. parahaemolyticus* and the dose-response relation are major sources of the uncertainties in the final result.

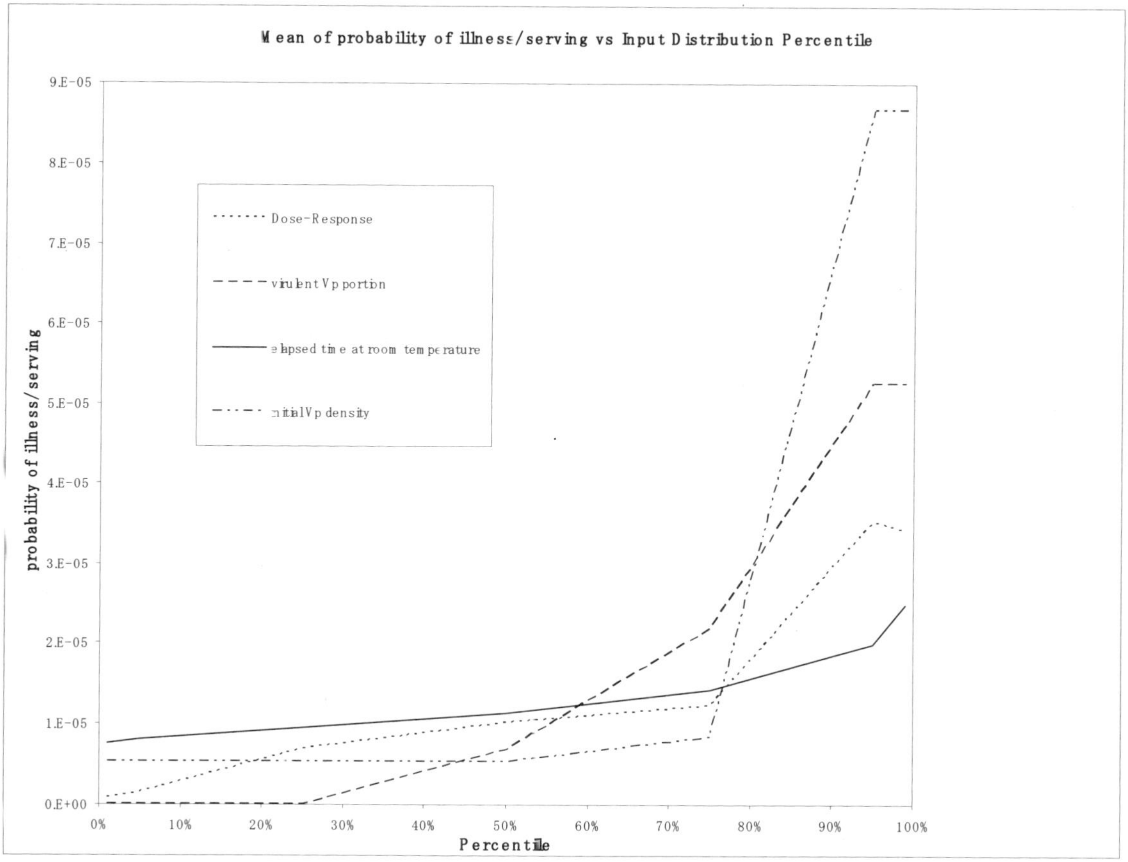

Figure III–11. Sensitivity analysis spider plot: x-axis is percentile of uncertainty parameter; y-axis is mean of the risk estimate.

III–7. Discussions and Conclusions

A risk assessment model of *V. parahaemolyticus* in horse mackerel was developed using data collected from published and unpublished literature, and data generated during the process of preparing this risk assessment. The probability of becoming ill per serving of raw horse mackerel was estimated to be from 8.77×10^{-7} (best scenario) to 3.75×10^{-5} (worst scenario). The probability of becoming ill in the worst-case scenario is about 40 times greater than that in the best-case scenario. Washing the eviscerated cavity of horse mackerel during the preparation of *sashimi* has a greater impact on the risk of illness than the use of disinfected water at landing places. In this risk assessment, the effect of using disinfected water at ports and during transportation is shown to be much less significant. However, the data obtained by the experimental models might not reflect the actual effect of using clean water for washing and transportation. The immersion in water in the experiment model might have had little influence to the concentration of *V. parahaemolyticus* in gills and intestines because the periods for which fish were immersed in clean water might have been too short to have an effect. Further experimental studies and more relevant data would be necessary to determine the true effect of the use of clean water at landing places and during transport.

If the results presented in Table III–14 are assumed to be valid from June to October, the period during which most cases of foodborne illness caused by *V. parahaemolyticus* occur in Japan, the risk of illness associated with the consumption of raw horse mackerel per year for a person ranges from 5.3×10^{-7} (best scenario) to 1.0×10^{-5} (worst scenario). Furthermore, if we apply these numbers to the Japanese population, the number of *V. parahaemolyticus* illnesses associated with consuming raw horse mackerel is 70 to 1300 per year.

This risk assessment highlighted that the percentage of pathogenic *V. parahaemolyticus* (*tdh* and/or *trh* positive) among the total *V. parahaemolyticus* population is the one of the important inputs needed to provide more useful and accurate scientific information for risk managers, and this deserves further experimental studies. In addition, more data on storage and transport temperatures would reduce the uncertainty regarding growth. Furthermore, concentrations of *V. parahaemolyticus* presented in the gill and intestine of fishes, which could not be eliminated by washing, were considered to be potentially important risk factors.

III–8. Limitations and caveats

This risk assessment is an example model based on a representative finfish species, the horse mackerel.

The dose-response model and the growth rate model that were used in the FDA-VPRA draft risk assessment of *V. parahaemolyticus* in shellfish (Anon., 2001a) were also used in this risk assessment. These parameters may differ for other seafoods. The growth rate could differ even in part of a single kind of food, e.g. on the surface or gills, or in the intestine of finfish. More laboratory and epidemiological studies are required to create a more realistic model of the dose-response relationship and growth of *V. parahaemolyticus* in horse mackerel.

It was assumed that the temperature of storage was low for most of the stages from harvest to consumption and this is probably applicable to horse mackerel distributed normally. However, a small fraction of horse mackerel is kept alive up to just before preparation in traditional Japanese hotels or restaurants. Sugiyama et al. (2002) demonstrated that MPN number of *V. parahaemolyticus* in water in live-fish tanks increased when the water temperature was above 19°C, and was significantly higher (more than 2400 MPN/100 mL water) when the temperature was above 22°C. Yamazaki, Yamaguchi and Noguchi (2001) indicated that *V. parahaemolyticus* was isolated from 50% (9 of 18 samples) of water from live-fish tanks, with the highest MPN being 430/100 mL at 22 C water temperature. In this case, if the water in the live-fish water tank is contaminated with *V. parahaemolyticus*, the increase of *V. parahaemolyticus* during transport and storage may become significant. In addition, the food is kept at room temperature for a short time after preparation. The effect of this phase was estimated based on rough assumptions in the present model.

Another important point is cross-contamination from one horse mackerel to another during preparation. This has not been addressed in the present model. It should be noted that this factor does not need to be considered explicitly when the dose-response relation is linear and only the total amount of *V. parahaemolyticus* consumed in the population is significant. In such cases, cross-contamination between horse mackerel does not change the total risk to the population. In the present model, the number of *V. parahaemolyticus* ingested per serving is small enough (less than 10 000) to allow linear approximation in the dose-response relation.

The effect of cross-contamination during preparation process, e.g. from horse mackerel to the hand of the food handler, then from the hand to other foods, or from horse mackerel to cutting board, then from the cutting board to other foods, was not taken into account in this model. This cross-contamination route could be important because *V. parahaemolyticus* might grow significantly on other ready-to-eat food contaminated with *V. parahaemolyticus* through cooking utensils and then kept at room temperature (Maki, 2005), whereas prepared *sashimi* is usually kept cool. However there were no data available to include this element in the risk assessment model.

This model could be applied to other finfish, but in doing so consideration needs to be given as to whether these other finfish are caught at different sea locations, by different fishing methods, or are processed and prepared in different ways.

In this risk assessment, the percentage of pathogenic *V. parahaemolyticus* in the total *V. parahaemolyticus* was selected randomly, based on the published reports (shown in Table III–3). However, this percentage could be different in other parts of the world, and in different seafood products.

One of difficulties in conducting a risk assessment for finfish is the complexity of the contaminated parts in a finfish. For example, *V. parahaemolyticus* contaminates the intestines, gills and skin surfaces. There also exists the complexity in the variety of serotypes and pathogenicity of *V. parahaemolyticus* and in its ecology. The efficacy of attachment of *V. parahaemolyticus* to the skin surface is different depending on the existence of mucus and scales on finfish. Furthermore, preparation methods also vary a lot. In this risk assessment, only consumption of raw finfish was considered. However, data from Thailand suggest that undercooked mackerel is also likely to be incriminated in foodborne disease. If data for each step from harvest to consumption are obtained for a local setting, then this risk assessment can be modified to cover the undercooked finfish.

III–9. References for Part III – Finfish

Anon[ymous]. 1988. Micro-organisms in Foods. ICMSF.

Anon. 1995. National Nutrition Survey, Japan. Ministry of Health and Welfare, Japan.

Anon. 1998a. National Nutrition Survey, Japan. Ministry of Health and Welfare, Japan.

Anon. 1998b. Outbreak of *Vibrio parahaemolyticus* infections associated with eating raw oysters --Pacific Northwest, 1997. *Morbidity and Mortality Weekly Report,* 47(22): 457–462.

Anon. 1999. Outbreak of *Vibrio parahaemolyticus* infection associated with eating raw oysters and clams harvested from Long Island Sound--Connecticut, New Jersey, and New York, 1998. *Morbidity and Mortality Weekly Report,* 48(3): 48–51.

Anon. 2000a. Ten-year summary of outbreaks due to *V. parahaemolyticus* in Japan (1989–1999). Ministry of Health and Welfare, Japan.

Anon. 2000b. Household income and expenditure survey, Japan. Ministry of Health and Welfare, Japan.

Anon. 2001a. Draft risk assessment on the public health impacts of *Vibrio parahaemolyticus* in raw molluscan shellfish. Prepared by the *Vibrio parahaemolyticus* Risk Assessment Task Force. Center for Food Safety and Applied Nutrition, U.S. Food and Drug Administration, Washington, DC, USA. Available at: http://vm.cfsan.fda.gov/~dms/vprisk.html (Accessed 23 July 2009).

Anon. 2001b. Fishery and aquaculture production statistics annual report [in Japanese]. Ministry of Agriculture, Forestry and Fisheries, Japan.

Anon. 2001c. Foodborne outbreak investigation report in 1999 [in Japanese]. Ministry of Health, Labour and Welfare, Japan.

Atthasampunna, P. 1974. *Vibrio parahaemolyticus* food poisoning in Thailand. pp. 21–26, *in*: T. Fujino, G. Sakaguchi, R. Sakazaki and Y. Takeda, (editors). International Symposium on *Vibrio parahaemolyticus*. Saikon Publishing Co., Tokyo, Japan.

Baffone, W., Pianetti, A., Bruscolini, F., Barbieri, E. & Citterio, B. 2000. Occurrence and expression of virulence-related properties of *Vibrio* species isolated from widely consumed seafood products. *International Journal of Food Microbiology*, 54: 9–18.

Baross, J.A., Hanus, F.J. & Morita, R.Y. 1975. Survival of human enteric and other sewage micro-organisms under simulated deep-sea conditions. *Applied Microbiology*, 30(2): 309–318.

Bean, N.H., Maloney, E.K., Potter, M.E., Korazemo, P., Ray, B., Taylor, J.P., Seigler, S. & Snowden, J. 1998. Crayfish: a newly recognized vehicle for vibrio infections. *Epidemiology and Infection*, 121(2): 269–273.

Beuchat, L.R. 1973. Interacting effects of pH, temperature, and salt concentration on growth and survival of *Vibrio parahaemolyticus*. Applied Microbiology, 25(5): 844–846.

Beuchat, L.R. & Worthington, R.E. 1976. Relationships between heat resistance and phospholipid fatty acid composition of *Vibrio parahaemolyticus*. Applied and Environmental Microbiology, 31(3): 389–394.

Blackstone, G.M., Nordstrom, J.L., Vickery, M.C.L., Bowen, M.C., Meyer, R.F. & DePaola, A. 2003. Detection of pathogenic *Vibrio parahaemolyticus* in oyster enrichments by real time PCR. *Journal of Microbiological Methods*, 53(2): 149–155.

Cabanillas-Beltrán, H., Llausás-Magaña, E., Romero, R., Espinoza, A., García-Gasca, A., Nishibuchi, M., Ishibashi, M. & Gomez-Gil, B. 2006. Outbreak of gastroenteritis caused by the pandemic *Vibrio parahaemolyticus* O3:K6 in Mexico. *FEMS Microbiol Letters*, 265(1): 76–80.

Chan, K.Y., Woo, M.L., Lam, L.Y. & French, G.L. 1989. *Vibrio parahaemolyticus* and other halophilic vibrios associated with seafood in Hong Kong. *Journal of Applied Bacteriology*, 66(1): 57–64.

Chowdhury, A., Mori, K., Nakano, Y., Ishibashi, M. & Nishibuchi, M. 2001. Study on the contamination of the imported seafood by *Vibrio parahaemolyticus* using genetic methods [in Japanese]. *Journal of the Japanese Society for Bacteriology*, 56(1): 323.

Chun, D., Yong Seol, S., Tak, R. & Kyu Park, C. 1972. Inhibitory effect of glycerin on *Vibrio parahaemolyticus* and *Salmonella*. Applied Microbiology, 24(5): 675–678.

Cook, D.W., O'Leary, P., Hunsucker, J.C., Sloan, E.M., Bowers, J.C., Blodgett, R.J. & DePaola, A. 2002. *Vibrio vulnificus* and *Vibrio parahaemolyticus* in U.S. retail shell oysters: A national survey June 1998 to July 1999. *Journal of Food Protection*, 65: 79–87.

Daniels, N.A., MacKinnon, L., Bishop, R., Altekruse, S., Ray, B., Hammond, R.M., Thompson, S., Wilson, S., Bean, N.H., Griffin, P.M. & Slutsker, L. 2000. *Vibrio parahaemolyticus* infections in the United States, 1973–1998. *Journal of Infectious Diseases*, 181(5): 1661–1666.

DePaola, A., Hopkins, L.H., Peeler, J.T., Wentz, B. & McPhearson, R.M. 1990. Incidence of *Vibrio parahaemolyticus* in U.S. coastal waters and oysters. *Applied and Environmental Microbiology*, 56: 2299–2302.

El-Sahn, M.A., El-Banna, A.A. & El-Tabey Shehata, A.M. 1982. Occurrence of *Vibrio parahaemolyticus* in selected marine invertebrates, sediment, and seawater around Alexandria, Egypt. *Canadian Journal of Microbiology*, 28(11): 1261–1264.

Fang, S.W., Huang, W.W. & Chen, L.H. 1987. Contamination of seafood by *Vibrio parahaemolyticus* in Taiwan. *Zhonghua Min Guo Wei Sheng Wu Ji Mian Yi Xue Za Zhi*, 20(2): 140–147.

FAO. 2000. World trade in mackerel and horse mackerel. *Globefish Market Research Programme*, Vol. 64. 98 p.

FAO/WHO. 2002. Risk assessment of *Campylobacter* spp. in broiler chickens and *Vibrio* spp. in seafood. Report of a Joint FAO/WHO Expert Consultation, Bangkok, Thailand, 5–9 August 2002. *FAO Food and Nutrition Paper*, No. 75. Available at: http://www.fao.org/docrep/008/y8145e/y8145e00.htm

FDA [US Food and Drug Administration]. 2005. Quantitative Risk Assessment on the Public Health Impact of Pathogenic *Vibrio parahaemolyticus* in Raw Oysters. U.S. Department of Health and Human Services, U.S. Food and Drug Administration. See: http://www.fda.gov/Food/ScienceResearch/ResearchAreas/RiskAssessmentSafetyAssessment/ucm185746.htm

Gyobu, Y., Shima, T., Tanaka, D., Kimata, K., Isobe, J., Katori, K., Watahiki, M. & Nagai, Y. 2004. Detection of tdh gene of *Vibrio parahaemolyticus* from sea fish [in Japanese]. *Annual Report of Toyama Institute of Health*, 27: 121–123.

Hackney, C.R. & Dicharry, A. 1988. Seafood-borne bacterial pathogens of marine origin. Source: *Food Technology*, 42(3): 104–109.

Haddock, R.L. & Cabanero, A.F. 1994. The origin of non-outbreak *Vibrio parahaemolyticus* infections on Guam. *Tropical and Geographical Medicine*, 46(1): 42–43.

Hara-Kudo, Y., Nishina, T., Sugiyama, K., Saitoh, A., Nakagawa, H., Ichihara, T., Konuma, H., Hasegawa, J. & Kumagai, S. 2001. Detection of TDH producing *Vibrio parahaemolyticus* O3:K6 from naturally contaminated shellfish with an immunomagnetic separation method and a chromogenic agar medium [In Japanese]. *Kansenshogaku Zasshi*, 75: 955–960.

Hara-Kudo, Y., Sugiyama, K., Nishibuchi, M., Chowdhury, A., Yatsuyanagi, J., Ohtomo, Y., Saito, A., Nagano, H., Nishina, T., Nakagawa, H., Konuma, H., Miyahara, M. & Kumagai, S. 2003. Prevalence of pandemic thermostable direct hemolysin-producing *Vibrio parahaemolyticus* O3:K6 in seafood and the coastal environment in Japan. *Applied Environmental Microbiology*, 69(7): 3883–3891.

Hiro, Y., Yamamoto, T., Nishiki, K., Shoji, T., Takamura, Y., Takeuchi, M., Nakai, Y. & Asai, R. 1996. Experimental test of *Vibrio parahaemolyticus* contamination in shellfish [in japanese]. Proceedings of the National Conference of Food Safety Inspection. pp. 22–29.

Honda. T., Ni, Y.X. & Miwatani, T. 1988. Purification and characterization of a hemolysin produced by a clinical isolate of Kanagawa phenomenon-negative *Vibrio parahaemolyticus* and related to the thermostable direct hemolysin. *Infection and Immunity*, 56: 961–965.

IASR [Infectious Agents Surveillance Report]. No date. Recent report on isolations of bacteria from prefecture public health institutes and local health canters. Data available from the IASR Web site: http://idsc.nih.go.jp/iasr/index.html (Accessed 23 July 2009).

Iawashita, M. 1991. Situation of fish transportation and contamination of *Vibrio parahaemolyticus*. Proceedings of the National Conference of Food Safety Inspection, Japan.

Islam, M.S., Tasmin, R., Khan, S.I., Bakht, H.B.M., Mahmood, Z.H., Rahman, M.Z., Bhuiyan, N.A., Nishibuchi, M., Nair, G.B., Sack, R.B., Huq, A., Colwell, R.R. & Sack, D.A. 2004. Pandemic strains of O3:K6 *Vibrio parahaemolyticus* in the aquatic environment of Bangladesh. *Canadian Journal of Microbiology*, 50(10): 827–834

Jaksic, S., Uhitil, S., Petrak, T, Bazulic, D. & Karolyi, L.G. 2002. Occurrence of *Vibrio* spp. in sea fish, shrimps and bivalve molluscs from the Adriatic Sea. *Food Control*, 13: 491–493.

Kamruzzaman, M., Mashita (Matsumoto) C., Nishibuchi, M., 2006. Analysis of specific characteristics of pandemic clones of *Vibrio parahaemolyticus* [in Japanese]. *Journal of Japanese Society for Bacteriology*, 61(1): 122.

Kaneko, T. & Colwell, R.R. 1975. Incidence of *Vibrio parahaemolyticus* in Chesapeake Bay. *Applied Microbiology*, 30(2): 251–257.

Kaufman, G.E., Bej, A.K., Bowers, J. & DePaola, A. 2003. Oyster-to-oyster variability in levels of *Vibrio parahaemolyticus*. *Journal of Food Protection*, 66: 125–129.

Kiiyukia, C., Venkateswaran, K., Navarro, I.M., Nakano, H., Kawakami, H. & Hashimoto, H. 1989. Seasonal distribution of *Vibrio parahaemolyticus* serotypes along the oyster beds in Hiroshima coast. *Journal of the Faculty of Applied Biological Science*, 28: 49–61.

Koga, T., Hirota, N. & Takumi, K. 1999. Bactericidal activities of essential oils of basil and sage against a range of bacteria and the effect of these essential oils on *Vibrio parahaemolyticus*. *Microbiological Research*, 154(3): 267–273.

Kumagai, S., Hara-Kudo, Y., Miwa, N., Masuda, T., Konuma, H., Hasegawa, J. & Nishina, T. 2003. Effect of washing of finfish on *Vibrio parahaemolyticus* contamination, Report for the WHO Agreement of Performance of Work.

Kumazawa, N.H., Hori, K., Fujimori, K., Iwade, Y. & Sugiyama, A. 1999. Geographical features of estuaries for neritid gastropods including *Clithon retropictus* to preserve thermostable direct hemolysin-producing *Vibrio parahaemolyticus*. *Journal of Veterinary Medical Science*, 61(6): 721–724.

McCarter, L.L. & Wright, M.E. 1993. Identification of genes encoding components of the swarmer cell flagellar motor and propeller and a sigma factor controlling differentiation of *Vibrio parahaemolyticus*. *Journal of Bacteriology*, 175: 3361–3371.

McCarthy, S.A., DePaola, A., Kaysner, C.A., Hill, W.E. & Cook, D.W. 2000. Evaluation of non-isotopic DNA hybridization methods for detection of the *tdh* gene of *Vibrio parahaemolyticus*. *Journal of Food Protection*, 63: 1660–1664.

Maki, Y. 2005. *Vibrio parahaemolyticus* food poisoning affected by flood [in Japanese]. *Journal of the Food Hygienics Society of Japan*, 46(5): 299–300.

Makino, K., Oshima, K., Kurokawa, K., Yokoyama, K., Uda, T., Tagamori, K., Iijima, Y., Najima, M., Nakano, M., Yamashita, A., Kubota, Y., Kimura, S., Yasunaga, T., Honda, T., Shinagawa, H., Hattori, M. & Iida, T. 2003. Genome sequence of *Vibrio parahaemolyticus*: a pathogenic mechanism distinct from that of *V. cholera*. *Lancet*, 361: 743–749.

Martinez-Urtaza, J., Simental, L., Velasco, D., DePaola, A., Ishibashi,M., Nakaguchi, Y., Nishibuchi, M., Carrera-Flores, D., Rey-Alvarez, C. & Pousa, A. 2005. Pandemic *Vibrio parahaemolyticus* O3:K6, Europe. *Emerging Infectious Diseases*, 11(8): 1319–1320.

Matsumoto, C., Okuda, J., Ishibashi, M., Iwanaga, M., Garg, P., Rammamurthy, T., Wong, H.C., Depaola, A., Kim, Y.B., Albert, M.J. & Nishibuchi, M. 2000. Pandemic spread of an O3:K6 clone of *Vibrio parahaemolyticus* and emergence of related strains evidenced by arbitrarily primed PCR and toxRS sequence analyses. *Journal of Clinical Microbiology*, 38(2): 578–585.

Matte, G.R., Matte, M.H., Sato, M.I., Sanchez, P.S., Rivera, I.G. & Martins, M.T. 1994. Potentially pathogenic vibrios associated with mussels from a tropical region on the Atlantic coast of Brazil. *Journal of Applied Bacteriology*, 77(3): 281–287.

Miles, D.W., Ross, T., Olley, J. & McMeekin, T A, 1997. Development and evaluation of a predictive model for the effect of temperature and water activity on the growth rate of *Vibrio parahaemolyticus*. *International Journal of Food Microbiology*, 38(2–3): 133–142.

Molero, X., Bartolome, R.M., Vinuesa, T., Guarner, L., Accarino, A., Casellas, F. & Garcia, R. 1989. Acute *Vibrio parahaemolyticus* gastroenteritis in Spain - report of 8 cases. *Medicina Clinica (Barcelona)*, 92(1): 1–4.

Muntada-Garriga, J.M., Rodriguez-Jerez, J.J., Lopez-Sabater, E.I. & Mora-Ventura, M.T. 1995. Effect of chill and freezing temperatures on survival of *Vibrio parahaemolyticus* inoculated in homogenates of oyster meat. *Letters in Applied Microbiology,* 20(4): 225–227.

Nishibuchi, M., Kumagai, K. & Kaper, J.B. 1991. Contribution of the *tdh*1 gene of Kanagawa phenomenon-positive *Vibrio parahaemolyticus* to production of extracellular thermostable direct hemolysin. *Microbial Patholgenesis,* 11(6): 453–460.

Nishibuchi, M. & Kaper, J.B. 1995. Thermostable direct hemolysin gene of *Vibrio parahaemolyticus*: a virulence gene acquired by a marine bacterium. *Infection and Immunity,* 63: 2093–2099.

Ng, T.C., Chiang, P.C., Wu, T.L. & Leu, H.S. 1999. *Vibrio parahaemolyticus* bacteremia: case report. *Changgeng Yi Xue Za Zhi,* 22: 508–514.

Nordstrom, J.L., DePaola, A., Bowers, J.C., Wells, J.G. & Cook, D.W. 2003. Seasonal abundance of total and pathogenic *Vibrio parahaemolyticus* in Alabama oysters. *Applied and Environmental Microbiology,* 69: 1521–1526.

Ogasawara, H. 2000. Study on the future prediction and prevention of *Vibrio parahaemolyticus* outbreaks. Study report on environmental health. Ministry of Health and Welfare, Japan. pp. 43–51.

Ogawa, H., Tokunou, H., Kishimoto, T., Fukuda, S., Umemura, K. & Takata, M. 1989. Ecology of *Vibrio parahaemolyticus* in Hiroshima Bay. *The Hiroshima Journal of Veterinary Medicine,* 4: 47–57.

Ohno, S., Tazawa, T., Kon, M., Uno, Y., Terao, M. & Goto, K. 1993. Contamination by *Vibrio parahaemolyticus* of fish landed at fishery markets in Niigata Prefecture [in Japanese. *Niigataken Eisei Kogai Kenkyusyo Nenpo [Annual Report of Niigata Prefectural Research Laboratory for Health and Environment],* 9: 77–82.

Okuda, J., Ishibashi, M., Hayashi, E., Nishino, T., Takeda, Y., Mukhopadhyary, A.K., Garg, S., Bhattacharya, S.K., Nair, B.G. & Nishibuchi, M. 1997. Emergence of a unique O3:K6 clone of *Vibrio parahaemolyticus* in Calcutta, India, and isolation of strains from the same clonal group from southeast Asian travellers arriving in Japan. *Journal of Clinical Microbiology,* 35: 3150–3155.

Otomo, Y. & Yatsuyanagi, J. 2003. The latest trend and characteristic of *Vibrio parahaemolyticus* in Japan –Focusing on the research in Tohoku district. *Japanese Journal of Food Microbiology,* 20(4), 161-164

Oliver, J.D. 1981. Lethal cold stress of *Vibrio vulnificus* in oysters. *Applied and Environmental Microbiology,* 41(3): 710–717.

Robert-Pillot, A., Guenole, A., Lesne, J., Delesmont, R., Fournier, J.M. & Quilici, M.L. 2004. Occurrence of the tdh and trh genes in *Vibrio parahaemolyticus* isolates from waters and raw shellfish collected in two French coastal areas and from seafood imported into France. *International Journal of Food Microbiology,* 91(3): 319–325.

Sakazaki, R. & Nakanishi, H. 1975. Control of food poisoning with *Vibrio parahaemolyticus* – fundamental consideration and its practice. *Shokuhin Eisei Kenkyu,* 27(6): 527–532.

Sanyal, S.C. & Sen, P.C. 1974. Human volunteer study on the pathogenicity of *Vibrio parahaemolyticus*, pp. 227–230, *in*: T. Fujino, G. Sakaguchi, R. Sakazaki and Y. Takeda (editors). International Symposium on *Vibrio parahaemolyticus*. Saikon Publishing Co. Ltd., Tokyo, Japan.

Sarkar, B.L., Nair, G.B. Banerjee, A.K. & Pal, S.C. 1985. Seasonal distribution of *Vibrio parahaemolyticus* in freshwater environs and in association with freshwater fishes in Calcutta. *Applied and Environmental Microbiology,* 49(1): 132–136.

Shiozawa, K., Kubota, T., Furuya, Y. ,Akahane, S., Aoki, K., Moriyama, C., Watanabe, S., Yamaguchi, H., Yamaguchi, H., Yamauchi, K., Ohmura, Y., Satake, Y., Tada, T., Niimi, H., Ishii, H., Ike, N., Atsumi, M., Matsubayashi, S., Uekuzu, S., Matsushita, N., Kanda, M., Ikehata, A., Iida, T. & Murasawa, K. 1998. Study on the prediction of food poisoning due to *Vibrio parahaemolyticus*. II. Relationship between the viable cell number of *V. parahaemolyticus* in the sea water used to maintain short-necked clam and the number of food poisoning cases [in Japanese]. *In:* Abstract Book of the 28th Meeting of Shizuoka Prefecture Public Health Study Group. pp. 2-51 – 2-53.

Sugiyama, A., Nakano, Y., Iwade, Y., Yano, T., Fukuda, M., Yamauchi, A., Kawade, K., Sakurai, Y., Matsumoto, T. & Yamanaka, O. 2002. *Vibrio parahaemolyticus* food poisonings and their control measures based on HACCP principles. Report on food safety assurance project for agriculture and fishery products [in Japanese]. Mie Prefectural Science and Technology Promotion Center, 37–47.

Twedt, R., Spaulding, P. & Hall, H. 1969. Morphological, cultural, biochemical, and serological comparison of Japanese strains of *Vibrio parahaemolyticus* with related cultures isolated in the United States. *Journal of Bacteriology*, 98(2): 511–518.

Tangkanakul, W., Tharmaphornpilas, P., Datapon, D. & Sutantayawalee, S. 2000. Food poisoning outbreak from contaminated fish-balls. *Journal of the Medical Association of Thailand*, 83(11): 1289–1295.

Vuddhakul, V., Chowdhury, A., Laohaprertthisan, V., Pungrasamee, P., Patararungrong, N., Thianmontri, P., Ishibashi, M., Matsumoto, C. & Nishibuchi, M. 2000. Isolation of a pandemic clone of a *Vibrio parahaemolyticus* strain from environmental and clinical sources in Thailand. Applied and Environmental Microbiology, 66: 2685–2689.

Wagatsuma, S. 1974. Ecological studies on Kanagawa phenomenon positive strains of *Vibrio parahaemolyticus*. pp. 91–96, *in:* T. Fujino, G. Sakaguchi, R. Sakazaki and Y. Takeda, (editors). International symposium on *Vibrio parahaemolyticus*. Saikon Publishing Co.,Tokyo, Japan.

Watanabe, T. 1994. Growth of *Vibrio parahaemolyticus* in different methods of cooking fish. Proceedings of National Conference of Food Safety Inspection, Japan.

Watkins, W.D. & Cabelli, V.J. 1985. Effect of fecal pollution on *Vibrio parahaemolyticus* densities in an estuarine environment. *Applied and Environmental Microbiology*, 49(5): 1307–1313.

Wong, H.C., Ting, S.H. & Shien, W.R. 1992. Incidence of toxigenic vibrios in foods available in Taiwan. *Journal of Applied Bacteriology*, 73: 197–202.

Wong H.C., Chen L.L. & Yu, C.M. 1995. Occurrence of vibrios in frozen seafoods and survival of *Vibrio cholerae* in broth and shrimp homogenate at low temperatures. *Journal of Food Protection*, 58: 263–267.

Wong, H.C., Chen, M.C., Liu, S.H. & Liu, D.P. 1999. Incidence of highly genetically diversified *Vibrio parahaemolyticus* in seafood imported from Asian countries. *International Journal of Food Microbiology*, 52(3): 181–188.

Yamai, S. 2001. A study of the contamination of *Vibrio parahaemolyticus*. Report of the research on environmental health, 2000. Ministry of Health and Welfare, Japan.

Yamamoto, S., Okujo, N., Yoshida, T., Matsuura, S. & Shinoda, S. 1994. Structure and iron transport activity of vibrioferrin, a new siderophore of *Vibrio parahaemolyticus*. *Journal of Biochemistry (Tokyo)*, 115: 868–874.

Yamamoto, S., Akiyama, T., Okujo, N., Matsu-ura, S. & Shinoda, S. 1995. Demonstration of a ferric vibrioferrin-binding protein in the outer membrane of *Vibrio parahaemolyticus*. *Microbiology and Immunology*, 39(10): 759–766.

Yamazaki, T., Masuno, H., Shimada, K., Ueno, E., Nakano, T., Kondo, Y., Hanji, T., Kojima, H., Okazaki, K., Nakajima, H. & Takakura, K. 1996. Detection of *Vibrio parahaemolyticus* from marine fish during a five-year study and detection of thermostable direct hemolysin (TDH) from the isolates [in Japanese]. Abstracts of the 1996 Meeting of National Workshop for Food Hygiene Inspectors, pp. 193–196.

Yamazaki, S., Yamaguchi, Y. & Noguchi, H. 2001. Investigation of *Vibrio parahaemolyticus* and the relation gene from marine products. *Bulletin of the Nagasaki Prefectural Institute for Environmental Research and Public Health*, 47: 112–114.

Yeung, P.S.M., Hayes, M.C., DePaola, A., Kaysner, C.A., Kornstein, L. & Boor, K.J. 2002. Comparative phenotypic, molecular, and virulence characterization of *Vibrio parahaemolyticus* O3:K6 isolates. *Applied and Environmental Microbiology*, 68(6): 2901–2909.

Appendix III–1

Ogawa et al. (1989) reported the relation between the concentration of *V. parahaemolyticus* in seawater n (number of *V. parahaemolyticus* in 100 mL of seawater); seawater temperature t (°C) and seawater salinity s (%).

The dependence on seawater temperature is given by:

$$\log_{10} n = 0.103\ t - 0.934.$$

The dependence on salinity is given by:

$$\log_{10} n = -0.674\ s + 3.448.$$

The dependence on seawater temperature and salinity is given by:

$$\log_{10} n = 2.00 + 0.05\ t - 0.584\ s \qquad (P<0.001)$$

The relationship between concentrations of *V. parahaemolyticus* in raw oyster and in seawater is:

$$\log_{10} n_{oy} = 0.900 \log_{10} n_{sw} + 1.232 \qquad (r=0.663, P<0.05)$$

where:

n_{oy} = number of *V. parahaemolyticus* in 100 g of raw oyster, and

n_{sw} = number of *V. parahaemolyticus* in 100 mL of seawater.

The concentration of *V. parahaemolyticus* in raw oyster was found to be 11.1 times greater than the concentration in seawater.

Reference cited

Ogawa, H., Tokunou, H., Kishimoto, T., Fukuda, S., Umemura, K. & Takata, M. 1989. Ecology of *Vibrio parahaemolyticus* in Hiroshima Bay. *The Hiroshima Journal of Veterinary Medicine*, 4: 47–57.

FAO/WHO MICROBIOLOGICAL RISK ASSESSMENT SERIES

1 Risk assessments of *Salmonella* in eggs and broiler chickens: Interpretative Summary, 2002.

2 Risk assessments of *Salmonella* in eggs and broiler chickens, 2002.

3 Hazard characterization for pathogens in food and water: Guidelines, 2004.

4 Risk assessment of *Listeria monocytogenes* in ready to eat foods: Interpretative Summary, 2004.

5 Risk assessment of *Listeria monocytogenes* in ready to eat foods: Technical Report, 2004.

6 *Enterobacter sakazakii* and other micro-organisms in powdered infant formula: Meeting Report, 2004.

7 Exposure assessment of microbiological hazards in food: Guidelines, 2008.

8 Risk assessment of *Vibrio vulnificus* in raw oysters: Interpretative Summary and Technical Report, 2005.

9 Risk assessment of choleragenic *Vibrio cholerae* O1 and O139 in warm-water shrimp in international trade: Interpretative Summary and Technical Report, 2005.

10 *Enterobacter sakazakii* and *Salmonella* in powdered infant formula: Meeting Report, 2006.

11 Risk assessment of *Campylobacter* spp. in broiler chickens: Interpretative Summary, 2009.

12 Risk assessment of *Campylobacter* spp. in broiler chickens: Technical Report, 2009.

13 Viruses in food: scientific advice to support risk management activities: Meeting Report, 2008.

14 Microbiological hazards in fresh leafy vegetables and herbs: Meeting Report, 2008.

15 *Enterobacter sakazakii* (*Cronobacter* spp.) in powdered follow-up formula: Meeting Report, 2008.

16 Risk assessment of *Vibrio parahaemolyticus* in seafood: Interpretative summary and Technical Report, 2011.

17 Risk characterization of microbiological hazards in food: Guidelines, 2009.

18 Enterohaemorrhagic *Escherichia coli* in raw beef and beef products: approaches for the provision of scientific advice: Meeting report, 2011.

19 *Salmonella* and *Campylobacter* in chicken meat: Meeting Report, 2009.